ASCE STANDARD ASCE/SEI 24-14

American Society of Civil Engineers

Flood Resistant Design and Construction

This document uses both the International System of Units (SI) and customary units.

AMERICAN SOCIETY OF CIVIL ENGINEERS

STRUCTURAL ENGINEERING INSTITUTE

Published by the American Society of Civil Engineers

Library of Congress Cataloging-in-Publication Data

Flood resistant design and construction.
pages cm
"ASCE/SEI 24-14."
ISBN 978-0-7844-1379-1 (print : alk. paper)–ISBN 978-0-7844-7872-1 (ebook)
1. Structural design–Standards. 2. Flood damage prevention–United States.
3. Building, Stormproof–United States. I. American Society of Civil Engineers.
TA658.2.F56 2015
624.1'76–dc23
2014035187

Published by American Society of Civil Engineers
1801 Alexander Bell Drive
Reston, Virginia, 20191-4382
www.asce.org/bookstore | ascelibrary.org

This standard was developed by a consensus standards development process that has been accredited by the American National Standards Institute (ANSI). Accreditation by ANSI, a voluntary accreditation body representing public and private sector standards development organizations in the United States and abroad, signifies that the standards development process used by ASCE has met the ANSI requirements for openness, balance, consensus, and due process.

While ASCE's process is designed to promote standards that reflect a fair and reasoned consensus among all interested participants, while preserving the public health, safety, and welfare that is paramount to its mission, it has not made an independent assessment of and does not warrant the accuracy, completeness, suitability, or utility of any information, apparatus, product, or process discussed herein. ASCE does not intend, nor should anyone interpret, ASCE's standards to replace the sound judgment of a competent professional, having knowledge and experience in the appropriate field(s) of practice, nor to substitute for the standard of care required of such professionals in interpreting and applying the contents of this standard.

ASCE has no authority to enforce compliance with its standards and does not undertake to certify products for compliance or to render any professional services to any person or entity.

ASCE disclaims any and all liability for any personal injury, property damage, financial loss, or other damages of any nature whatsoever, including without limitation any direct, indirect, special, exemplary, or consequential damages, resulting from any person's use of, or reliance on, this standard. Any individual who relies on this standard assumes full responsibility for such use.

Errata: Errata, if any, can be found at http://dx.doi.org/10.1061/9780784413791.

ISBN 978-0-7844-1379-1 (print)
ISBN 978-0-7844-7872-1 (PDF)
Manufactured in the United States of America.

28 27 26 25 24 4 5 6 7 8

ASCE STANDARDS

In 2006, the Board of Direction approved the revision to the ASCE Rules for Standards Committees to govern the writing and maintenance of standards developed by the Society. All such standards are developed by a consensus standards process managed by the Society's Codes and Standards Committee (CSC). The consensus process includes balloting by a balanced standards committee made up of Society members and nonmembers, balloting by the membership of the Society as a whole, and balloting by the public. All standards are updated or reaffirmed by the same process at intervals not exceeding five years.

The material presented in this Standard has been prepared in accordance with recognized engineering principles. This Standard should not be used without first securing competent advice with respect to its suitability for any given application. The publication of the material contained herein is not intended as a representation or warranty on the part of the American Society of Civil Engineers, or of any other person named herein, that this information is suitable for any general or particular use or promises freedom from infringement of any patent or patents. Anyone making use of this information assumes all liability from such use.

A complete list of currently available standards is available in the ASCE Library (http://ascelibrary.org/page/books/s-standards).

PREFACE

The material presented in this standard has been prepared in accordance with recognized engineering principles. This standard should not be used without first securing competent advice with respect to its suitability for any given application. The publication of the material contained herein is not intended as a representation or warranty on the part of the American Society of Civil Engineers, or of any other person named herein, that this information is suitable for any general or particular use or promises freedom from infringement of any patent or patents. Anyone making use of this information assumes all liability from such use.

The 2014 edition has a number of significant technical revisions from the 2005 edition, including

1. Defines Flood Design Class rather than using Risk/Occupancy Classification assigned under ASCE 7 and requires each building or structure governed by the standard to be assigned to Flood Design Class 1, 2, 3, or 4. Uses the assigned Flood Design Class to apply elevation requirements specified in Chapters 2, 4, 5, 6, and 7. Flood Design Class 4 buildings and facilities are equivalent to Occupancy Category/Risk Category IV buildings, which ASCE 7 identifies are essential facilities.
2. Adds definitions for Mixed Use and Residential Portions of Mixed Use in commentary to clarify limitations on use of dry floodproofing measures.
3. Changes the Coastal A Zone determination requirement from the designer's responsibility to one depending on either: (1) delineation of a Limit of Moderate Wave Action (LiMWA) on a Flood Insurance Rate Map, or (2) designation by the Authority Having Jurisdiction.
4. Separates specifications for flood openings from the installation requirements. Requires the presence of louvers, blades, screens, faceplates, or other covers and devices to be accounted for in determining net open area for non-engineered openings and in determining the performance of engineered openings. Revises coefficient of discharge table for engineered flood openings. Adds commentary regarding selection of coefficient of discharge and for grouping or stacking of flood openings.
5. For Flood Design Class 4 buildings, requires the minimum lowest floor elevation (or floodproofing level of protection) to be the higher of the Base Flood Elevation plus freeboard specified in Chapters 2, 4, and 6, the Design Flood Elevation, or the 500-year flood elevation. The 500-year flood elevation requirement is new.
6. Clarifies text pertaining to alluvial fan high risk flood hazard areas.
7. In Coastal High Hazard Areas (V Zone) and Coastal A Zones (if delineated),
    a. Makes explicit that designs must account for local scour and erosion
    b. Provides for shallow foundations in Coastal A Zones under certain circumstances
    c. Requires flood openings in breakaway walls
    d. Eliminates orientation of the lowest horizontal structural member as a factor to determine elevation for lowest floors, equipment, and flood damage-resistant materials
    e. Requires exterior doors at the top of stairways that are located inside enclosed areas with breakaway walls
    f. Consolidates requirements for all nonstructural concrete slabs
    g. Allows substantial improvement of existing buildings seaward of the reach of mean high tide in V zones (makes ASCE 24 consistent with National Flood Insurance Program and Coastal A Zones.
8. Updates flood damage-resistant material requirements.
9. Clarifies emergency escape and rescue opening requirements for dry floodproofed buildings.
10. Clarifies requirements for garages, carports, and accessory storage structures. Adds new section for multistory parking structures.
11. Consolidates requirements for tanks and more clearly distinguishes between requirements based on flood hazard area.

ACKNOWLEDGMENTS

The American Society of Civil Engineers (ASCE) acknowledges the work of the Flood Resistant Design and Construction Standard Committee of the Codes and Standards Activities Division of the Structural Engineering Institute. This group comprises individuals from many backgrounds including consulting engineering, research, construction, education, government, design, and private practice.

This standard was prepared through the consensus standards process by balloting in compliance with procedures of ASCE's Codes and Standards Activities Committee. Those individuals who served on the ASCE 24-14 Standard Committee include

Christopher P. Jones, P.E., M.ASCE, Chair
Larry Buss, P.E., D.WRE, M.ASCE, CFM
Russell J. Coco, P.E., M.ASCE
James P. Colgate, RA, Esq., CFM
William L. Coulbourne, P.E., M.ASCE, F.SEI
James B. Destefano, P.E., AIA, F.SEI
Gary J. Ehrlich, P.E., M.ASCE
Shou-Shan Fan, Ph.D.
Kenneth J. Filarski, FAIA, AICP, LEED-AP BD+C, CFM, SAP+AEER, NCARB
Daryle L. Fontenot, P.E., CFM
Carol Friedland, Ph.D., P.E., CFM
Michael J. Graham, CFM
John L. Ingargiola, EI, CBO, CFM
Catherine M. Kaake, P.E., M.ASCE
Philip Line, P.E., M.ASCE
E. Michael McCarthy, P.E., M.ASCE
Joseph J. Messersmith, Jr., P.E., M.ASCE
Kimberly Paarlberg, P.A.
John Squerciati, P.E., CFM, M.ASCE
Terri L. Turner, AICP, CFM
Robert A. Wessel, Ph.D., F.ASTM
Thomas G. Williamson, P.E., F.ASCE, F.SEI
Garland Wilson, P.E., M.ASCE

ASCE would like to thank each member of the committee for his or her dedication and hard work during completion of ASCE 24-14.

ASCE would like to acknowledge the long-standing contributions of previous Committee Chair, Harry B. Thomas, and of those current members who have served the committee since the development of the 1998 edition: William L. Coulbourne, Shou-Shan Fan, Christopher P. Jones, Joseph J. Messersmith, Jr., and Kimberly Paarlberg.

UNIT CONVERSIONS

Measurement	S.I. Units	Customary Units
Abbreviations	m = meter (S.I. base unit of length) cm = centimeter km = kilometer ha = hectare L = liter (S.I. base unit of volume) mL = milliliter kg = kilogram (S.I. base unit of mass) g = gram N = Newton ($m{\cdot}kg{\cdot}s^2$) Pa = Pascal (N/m^2) kPa = kilopascal J = Joule W = watt kW = kilowatt s = second (S.I. base unit of time) min = minute h = hour day °C = degrees Celsius ppm = parts per million	yd = yard in. = inch mi = mile acre gal = gallon qt = quart lb = pound oz = ounce lbf = pound-force (lb/ft) psi = pounds per square inch atm = atmosphere ft·lbf = feet per pound-force Btu = British thermal unit hp = horsepower s = second min = minute h = hour day °F = degrees Fahrenheit ppm = parts per million
Length	1 m = 3.2808 ft = 1.0936 yd 1 cm = 0.3937 in. 1 km = 0.6214 mile	1 ft = 0.333 yd = 0.3048 m 1 in. = 2.54 cm 1 mile = 0.869 nautical mile = 1.6093 km
Area	1 m^2 = 10.7643 ft^2 1 km^2 = 0.3861 mi^2 1 ha = 2.4710 acre	1 ft^2 = 0.0929 m^2 1 mi^2 = 2.59 km^2 1 acre = 43,560 ft^2 = 0.4047 ha
Volume	1 L = 0.2642 gal 1 ml = 1 cm^3	1 gal = 4 qt = 3.7854 L 1 ft^3 = 7.481 gal = 28.32 L
Mass	1 g = 0.0353 oz 1 kg = 2.2046 lb	1 oz = 28.3495 g 1 lb = 0.4536 kg
Force	1 N = 0.2248 lb/ft	1 lbf = 4.4482 N
Density	1 kg/m^2 = 0.2048 lb/ft^2 1 kg/m^3 = 6.2427 lb/ft^3	1 lb/ft^2 = 4.882 kg/m^2 1 lb/ft^3 = 16.018 kg/m^3
Pressure	1 kPa = 0.145 psi	1 psi = 6.8948 kPa 1 atm = 14.7 psi = 101.35 kPa
Energy and Power	1 J = 1.00 W·s = 0.7376 ft·lbf 1 kJ = 0.2778 W·h = 0.948 Btu 1 W = 0.7376 ft·lbf/s = 3.4122 Btu/h 1 kW = 1,3410 hp	1 ft·lbf = 1.3558 J 1 Btu = 1.0551 kJ 1 ft·lbf/s = 1.3558 W 1 hp = 550 ft·lb/s = 0.7457 kW
Flow	1 L/s = 15.85 gal/min = 2.119 ft^3/min	1 gal/min = 0.1337 ft^3/min = 0.0631 L/s
Concentration	mg/L = ppm_m (in dilute solutions)	
Temperature	°C = (°F − 32) × 5/9	°F = (°C × 9/5) + 32
Fundamental Constants and Relationships	Acceleration of gravity Density of freshwater (at 4 °C) = Density of saltwater (at 15 °C, 35 ppt) = Specific weight of freshwater (15 °C) = Specific weight of saltwater Weight of freshwater Weight of saltwater	32.2 ft/s^2 = 9.81 m/s^2 1,000 kg/m^3 = 1 g/cm^3 1,025 kg/m^3 = 1.025 g/cm^3 62.4 lb/ft^3 = 9,810 N/m^3 64.0 lb/ft^3 = 10,062 N/m^3 1 gal = 8.345 lb = 3.7854 kg 1 gal = 8.559 lb = 3.8825 kg

CONTENTS

Commentary

CHAPTER 1
GENERAL

1.1 SCOPE

This standard provides minimum requirements for flood resistant design and construction of structures that are subject to building code requirements and that are located, in whole or in part, in Flood Hazard Areas. This standard applies to the following: (1) new construction, including subsequent work to such structures, and (2) work classified as substantial improvement of an existing structure that is not an historic structure (see Fig. 1-1).

The general provisions of this section shall apply to all new construction and substantial improvements in flood hazard areas. In addition to the requirements of this section (see Fig. 1-2):

1. Chapter 2 shall apply to all new construction and substantial improvements in Flood Hazard Areas and High Risk Flood Hazard Areas except those that are identified as Coastal High Hazard Areas and Coastal A Zones;
2. Chapter 3 shall apply to all new construction and substantial improvements in High Risk Flood Hazard Areas;
3. Chapter 4 shall apply to all new construction and substantial improvements in Coastal High Hazard Areas and Coastal A Zones; and
4. Chapters 5, 6, 7, 8, and 9 shall apply to all new construction and substantial improvements.

1.2 DEFINITIONS

The following definitions apply to the provisions of the entire standard (italicized words in a definition mean the words are defined in this section):

500-Year Flood Elevation—Elevation of flooding having a 0.2% chance of being equaled or exceeded in any given year.

Accessory Storage Structure—A *structure* designed and used only for storage that is customarily accessory to and incidental to that of dwellings.

Alluvial Fan—Fan-shaped deposits of sediment eroded from steep slopes and *watersheds* and deposited on valley floors.

Alluvial Fan Flooding—Type of flood hazard that occurs only on *alluvial fans*. Alluvial fan flooding is considered hazardous when designated as a *flood hazard area* on a community's *flood hazard map* or otherwise legally designated.

Apex—Highest point on an *alluvial fan* or similar landform, where the flow is last confined. The apex generally corresponds to the location where the *watershed* erosion ceases and fan sediment deposition commences.

Attendant Utilities and Equipment—Utilities, mechanical, electrical, fuel gas, plumbing, HVAC, and related equipment, as well as services associated with new construction and *substantial improvements*.

Authority Having Jurisdiction—Organization, community, political subdivision, office, or agency that has adopted this standard under due legislative authority.

Base Flood—Flood having a 1% chance of being equaled or exceeded in any given year.

Base Flood Elevation (BFE)—Elevation of flooding, including *wave height*, having a 1% chance of being equaled or exceeded in any given year.

Basement—That portion of a *structure* having its lowest floor below ground level on all sides.

Breakaway Wall—Any type of wall subject to flooding that is not required to provide structural support to a building or other *structure* and that is designed and constructed such that, under *base flood* or lesser flood conditions, it will collapse under specific lateral loads in such a way that (1) it allows the free passage of floodwaters, and (2) it does not damage the structure or supporting foundation system.

Bulkhead—Wall or *structure* to retain or prevent sliding or *erosion* of the land; sometimes used to protect against *wave* action.

Channel—Natural or artificial waterway that periodically or continuously contains moving water.

Coastal A Zone (CAZ)—Area within a *special flood hazard area*, landward of a *V Zone* or landward of an open coast without mapped *V Zones*. In a Coastal A Zone, the principal source of flooding must be astronomical tides, storm surges, seiches, or tsunamis, not riverine flooding. During the *base flood* conditions, the potential for breaking *wave heights* shall be greater than or equal to 1.5 ft. The inland limit of the Coastal A Zone is (1) the *Limit of Moderate Wave Action* if delineated on a *FIRM*, or (2) designated by the *authority having jurisdiction*.

Coastal High Hazard Area (CHHA)—Area within a *special flood hazard area* extending from offshore to the inland limit of a primary frontal dune along an open coast and any other area that is subject to *high velocity wave action* from storms or seismic sources. This area is designated on *FIRM*s as velocity zones V, VO, VE, or V1-30.

Community—Any state or area or political subdivision thereof, or any Indian tribe or authorized tribal organization, or Alaska native village or authorized native organization, which has the authority to adopt and enforce this standard for areas within its jurisdiction.

Datum—The vertical reference on which maps are drawn, including but not limited to the North American Vertical Datum of 1988 (NAVD) and the National Geodetic Vertical Datum of 1929 (NGVD).

Debris Flow—Mass movement of sediment, including boulders, organic materials, and other debris; debris flows typically move in surges and are characterized by a steep frontal wave.

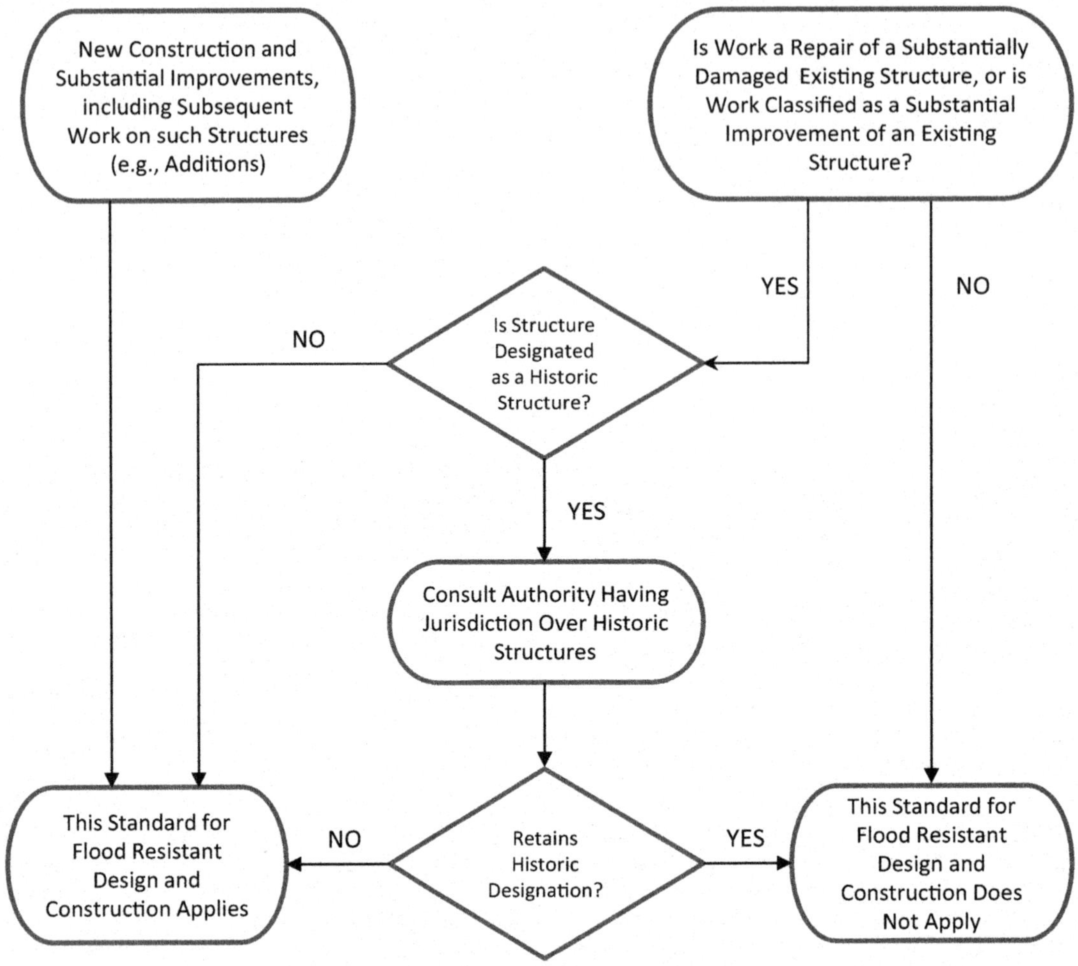

Fig. 1-1. Illustration of application of this standard.

Debris Impact Loads—Loads on a *structure* caused by flood-borne debris striking the structure, or a portion thereof; often it is sudden in nature and large in magnitude.

Design Flood—The flood associated with the greater of the following two areas: (1) area within a *floodplain* subject to a 1% or greater chance of flooding in any year, or (2) area designated as a *flood hazard area* on a *community's flood hazard map* or otherwise legally designated.

Design Flood Elevation (DFE)—Elevation of the *design flood*, including *wave height*, relative to the *datum* specified on the community's *flood hazard map*.

Dry Floodproofing—A combination of measures that results in a *structure*, including the *attendant utilities and equipment*, being watertight with all elements *substantially impermeable* and with structural components having the capacity to resist flood loads.

Enclosed Area or Enclosure—Confined area below the *DFE*, formed by walls on all sides of the enclosed space.

Eroded Ground Elevation—Elevation of ground following *erosion*.

Erodible Soil—Soil subject to wearing away, shifting, and movement due to the effects of wind, water, or other geological processes during a flood or storm or over a period of years.

Erosion—Wearing away of the land surface by detachment and movement of soil and rock fragments during a flood or storm or over a period of years, through the action of wind, water, or other geologic processes.

Erosion Analysis—Analysis of the *erosion* potential of soil or strata to include the effects of *flooding* or storm surge, moving water, *wave* action, and the interaction of water and structural components.

Essential Facility—Buildings and other *structures* that are intended to remain operational in the event of extreme environmental loading from flood, wind, snow, or earthquakes.

Existing Structure—Any *structure* for which the *start of construction* commenced before the effective date of the first *floodplain* management code, ordinance, or standard adopted by the *authority having jurisdiction*.

Fill—Material such as soil, gravel, or crushed stone that is placed in an area to increase ground elevations (see *Structural Fill*).

Flash Flood—*Flood* that crests in a short length of time and is often characterized by *high velocity flow*; it often results from heavy rainfall over a localized area, which overflows a confined water course. A flood whose waters rise from within banks to 3 ft or more above banks in less than 2 hours shall be considered a flash flood.

Flood or Flooding—General and temporary condition of partial or complete inundation of normally dry land from (1) the overflow of inland or tidal waters, or (2) the unusual and rapid accumulation of runoff or surface waters from any source.

Flood Control Structure—Barrier designed and constructed to keep water away from or out of a specified area (see *Flood Protective Works*).

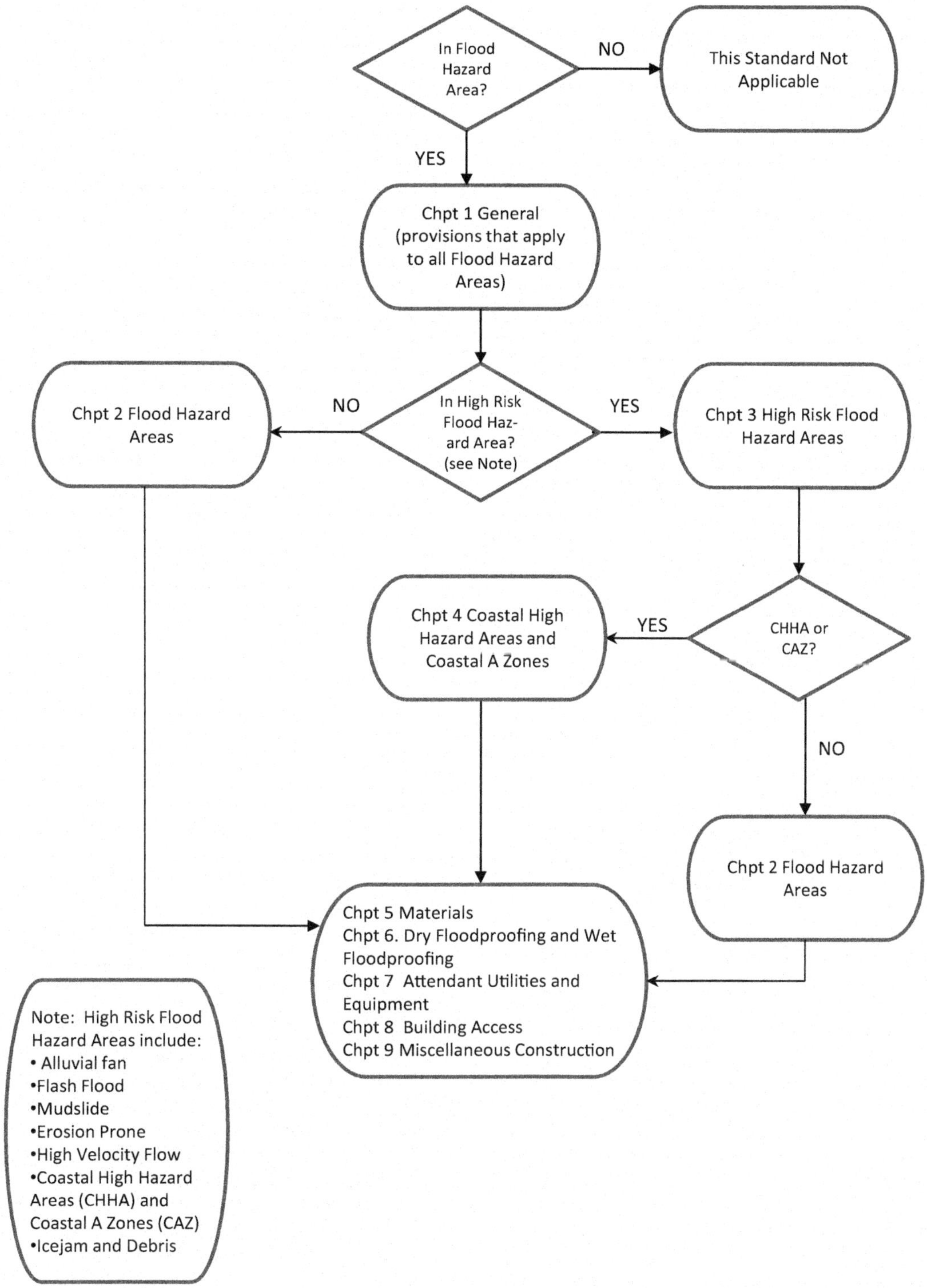

Fig. 1-2. Application of chapters.

Flood Damage-Resistant Material—Any construction material capable of withstanding direct and *prolonged contact with floodwaters* without sustaining any damage that requires more than cosmetic repair.

Flood Design Class–A classification of buildings and other structures for determination of flood loads and conditions, and determination of minimum elevation requirements on the basis of risk associated with unacceptable performance.

Flood Hazard Area—Area subject to *flooding* during the *design flood.*

Flood Hazard Map—Map delineating *flood hazard areas* adopted by the *authority having jurisdiction.*

Flood Hazard Study—Study that serves as the technical basis for a *flood hazard map.*

Flood Insurance Rate Map (FIRM)—Official map of a *community* on which the Federal Emergency Management Agency

(FEMA) has delineated both *special flood hazard areas* and the risk premium zones applicable to the community.

Flood Protective Works—Barriers designed and constructed to keep water away from or out of a specified area (see *Flood Control Structure; Floodwall*).

Flood-Related Erosion—Collapse, subsidence, or wearing away of land as a result of the action of *flooding*, including the effects of storm surge, moving water, and *wave* action.

Floodplain—Any land area, including watercourse, susceptible to partial or complete inundation by water from any source.

Floodproofing—Any combination of structural or nonstructural adjustments, changes, or actions that reduce or eliminate flood damage to a *structure*, contents, and *attendant utilities and equipment*.

Floodwall—Constructed barrier used to contain, control, or divert the flow of water so as to reduce the frequency of *flooding* in a specific area.

Floodway—Channel and that portion of the *floodplain* reserved to convey the base flood without cumulatively increasing the water surface elevation more than a designated height.

Footing—Enlarged base of a foundation, wall, pier, or column designed to spread the load of the *structure* so that it does not exceed the soil-bearing capacity.

Footprint—Horizontal extent of a *structure*.

Functionally Dependent—A use which cannot perform its intended purpose unless it is located or carried out in close proximity to water. The term includes only docking facilities, port facilities that are necessary for the loading and unloading of cargo or passengers, and ship building and ship repair facilities, but does not include long-term storage or related manufacturing facilities.

High Risk Flood Hazard Area—*Flood hazard area* where one or more of the following hazards are known to occur: *alluvial fan flooding, flash floods, mudslides, ice jams, high velocity flows, high velocity wave action,* breaking wave heights greater than or equal to 1.5 ft (*Coastal High Hazard Area* and *Coastal A Zone*), or *erosion*.

High Velocity Flow—During *design flood* or lesser conditions, water movement adjacent to *structures* or foundations with flow velocities greater than 10 ft/s.

High Velocity Wave Action—Condition where *wave heights* are greater than or equal to 3.0 ft or where *wave runup elevations* reach 3.0 ft or more above grade.

Highest Adjacent Grade—Highest elevation of the natural ground surface prior to construction next to the proposed foundation of a *structure*.

Historic Structure—Any *structure* that meets one of the following criteria: (1) listed individually in the National Register of Historic Places; (2) certified by the Secretary of the Interior as meeting the requirements for individual listing in the National Register; (3) certified or preliminary determination by the Secretary of the Interior as contributing to the historical significance of a registered historic district or a district preliminarily determined by the Secretary of the Interior to qualify as a registered historic district; (4) individually listed on a state inventory of historic places, in states with historic preservation programs that have been approved by the Secretary of the Interior; or (5) individually listed on a local inventory of historic places in *communities* with historic programs certified by an approved state program or by the Secretary of the Interior.

Human Intervention—Required presence and active involvement of people to implement a *floodproofing* measure prior to the onset of *flooding*.

Hydrodynamic Loads—Loads imposed on an object by water flowing against and around it.

Hydrostatic Loads—Loads imposed on an object by a standing mass of water.

Ice Jam—Accumulation of floating ice fragments that causes the bridging or damming of a channel or stream.

Impact Loads—Loads that result from debris, ice, or any object transported by floodwaters striking against *structures* or parts thereof.

Levee—Artificial barrier, usually an earthen embankment, used to contain, control, or divert the flow of water so as to reduce the frequency of *flooding* in a specific area.

Limit of Moderate Wave Action (LiMWA)—Line shown on *FIRMs* to indicate the inland limit of the 1.5-ft breaking wave height during the base *flood*.

Local Scour—During flood conditions, the removal of material from a localized portion of the channel cross section or land surface due to an abrupt change in flow direction or velocity around an object or structural element.

Lowest Floor—Lowest floor of the lowest enclosed area, including *basement*; however, an unfinished or flood-resistant enclosure used solely for parking of vehicles, building access, or storage shall not be considered the lowest floor provided such enclosure is built as specified in this standard.

Mangrove Stand—Assemblage of mangrove trees containing one or more of the following species: black mangrove, red mangrove, white mangrove, or buttonwood.

Mud Flood—Hyperconcentrated sediment flow with a sediment concentration less than 45% by volume. Characterized by distinct fluid properties in deformation, particle setting, wave motion, and spreading on a horizontal surface. A mud flood is a turbulent *flood* phenomenon.

Mudflow—Hyperconcentrated sediment flow with a sediment concentration in excess of 45% by volume, characterized by plastic deformation, cohesive characteristics, and lack of *wave* motion and spreading on a horizontal surface. A mudflow is a viscous flow.

Mudslide—General category of hyperconcentrated sediment flows including *mudflows, mud floods,* and *debris flows*.

New Construction—Structures for which the *start of construction* commenced on or after the effective date of the first *floodplain* management code, regulation, ordinance, or standard adopted by the *authority having jurisdiction*, including any subsequent improvements to such structures. New construction includes work determined to be *substantial improvement*.

Nonerodible Soil—Soil not subject to wearing away or movement due to the effects of wind, water, or other agents during a flood or storm or a period of years.

Nonresidential—Any building or *structure* or portion thereof that is not classified *residential*.

Obstruction—Any object or structural component attached to a *structure* below the *DFE* that can cause an increase in *flood* elevation, deflect floodwaters, or transfer flood loads to any structure.

Pile—Structural element that is embedded in soils by drilling, driving, or jetting, so that axial loads in the member are supported through skin friction or end bearing with the soil or rock and lateral loads in the member are supported through side bearing with the soil or rock.

Prolonged Contact with Floodwaters—Partial or total inundation by floodwaters for 72 h or more.

Rapid Drawdown—Rapid lowering of flood elevation at a rate equal to or exceeding 5 ft/h.

Rapid Rise—Rapid increase in flood elevation at a rate equal to or exceeding 5 ft/h.

Residential—(1) Buildings and structures and portions thereof where people live or that are used for sleeping purposes on a transient or nontransient basis; (2) *structures* including but not limited to one- and two-family dwellings, townhouses, condominiums, multifamily dwellings, apartments, congregate residences, boarding houses, lodging houses, rooming houses, hotels, motels, apartment buildings, convents, monasteries, dormitories, fraternity houses, sorority houses, vacation time-share properties; and (3) institutional facilities where people are cared for or live on a 24-h basis in a supervised environment, including but not limited to board and care facilities, assisted living facilities, halfway houses, group homes, congregate care facilities, social rehabilitation facilities, alcohol and drug centers, convalescent facilities, hospitals, nursing homes, mental hospitals, detoxification facilities, prisons, jails, reformatories, detention centers, correctional centers, and prerelease centers.

Sand Dune—Natural or artificial ridges or mounds of sand landward of a beach.

Seawall—Wall separating land and water areas, primarily designed to prevent *erosion* and other damage due to *wave* action.

Shear Wall—Load bearing or nonload–bearing wall that transfers, by in-plane lateral forces, lateral loads acting on a *structure* to its foundation.

Shield—Removable or permanent substantially impermeable protective cover for an opening in a *structure* below the *DFE*, used in *dry floodproofing* the *structure*.

Special Flood Hazard Area—Land in the *floodplain* subject to a 1% or greater chance of *flooding* in any given year; area delineated on the *Flood Insurance Rate Map* as Zone A, AE, A1-30, A99, AR, AO, AH, V, VO, VE, or V1-30.

Start of Construction—Date the construction permit was issued for *new construction*, provided that actual start of construction commenced within 180 days of the permit date. The actual start means either the first placement of permanent construction of a *structure* on a site, such as the pouring of a slab or *footing*, the installation of *piles*, the construction of columns, or any other work beyond the stage of excavation; or the placement of a manufactured home. Permanent construction does not include land preparation, such as clearing, grading, or filling; nor does it include excavation for a *basement, footings,* piers, or foundation or the erection of temporary forms; nor does it include the installation of accessory structures, such as garages or sheds not occupied as dwelling units or not part of the main structure. For *substantial improvement*, the actual start of construction means the first alteration of any wall, ceiling, floor, or other structural part of a *structure*, whether or not that alteration affects the external dimensions of the *structure*.

Stem Walls—Masonry or concrete perimeter walls backfilled with compacted soil or gravel to support a floor slab or floor system.

Stillwater Depth—Vertical distance between the ground and the *stillwater elevation*.

Stillwater Elevation—Elevation that the surface of the water would assume in the absence of waves referenced to a *datum*.

Storage Tank—Closed vessel used to store gases or liquids.

Structural Fill—*Fill* placed and compacted to a specified density to provide structural support or protection to a *structure*.

Structure—Any building or other *structure*, including gas and liquid *storage tanks*.

Substantial Damage—Damage of any origin sustained by a *structure*, whereby the cost of restoration to its predamage condition equals or exceeds 50% of its predamage market value, or equals or exceeds a smaller percentage established by the *authority having jurisdiction*.

Substantial Improvement—Any reconstruction, rehabilitation, addition, or other improvement to a *structure*, the cost of which equals or exceeds 50% of its preimprovement market value, or equals or exceeds a smaller percentage established by the *authority having jurisdiction*. This term includes structures that have incurred *substantial damage*, regardless of the actual repair work performed.

Substantially Impermeable—Use of *flood damage-resistant materials* and techniques for *dry floodproofing* portions of a *structure*, which result in a space free of through cracks, openings, or other channels that permit unobstructed passage of water and seepage during flooding, and which result in a maximum accumulation of 4 in. of water depth in such space during a period of 24 h.

V Zone—Velocity Zones V, VO, VE, or V1-30 (See *Coastal High Hazard Area*).

Watershed—Topographically defined area drained by a river or stream, or by a system of connecting rivers and streams such that all outflow is discharged through a single outlet.

Wave—Ridge, deformation, or undulation of the water surface.

Wave Height—Vertical distance between the crest and the trough of a *wave*.

Wave Loads—Loads imparted on a *structure* caused by *waves* striking the *structure* or a portion thereof.

Wave Runup—Rush of *wave* water running up a slope or *structure*.

Wave Runup Elevation—Elevation, usually referenced to a *datum*, reached by *wave runup*.

Wet Floodproofing—*Floodproofing* method that relies on the use of *flood damage-resistant materials* and construction techniques in areas of a *structure* that are below the elevation required by this standard by intentionally allowing those areas to flood (see *Floodproofing*).

1.3 IDENTIFICATION OF FLOOD HAZARD AREAS

This standard shall apply to the larger of (1) those lands within a floodplain subject to a 1% or greater chance of flooding in any year (i.e., the area subject to flooding during the base flood event); and (2) those lands designated as a flood hazard area on a community's flood hazard map, or otherwise legally designated.

The flood associated with the governing definition listed here shall be termed the design flood. Design and construction requirements for new construction and substantial improvements shall be dictated by conditions during the design flood.

1.4 IDENTIFICATION OF FLOOD-PRONE STRUCTURES

1.4.1 General A determination shall be made as to whether or not a structure lies, in whole or in part, within a flood hazard area following review of flood hazard maps, studies available in the public domain, and other information available from the authority having jurisdiction.

1.4.2 Consideration for Flood Protective Works Dams, levees, floodwalls, diversions, channels, flood control structures, and other flood protective works shall not be considered to provide protection for structures during the design flood, unless those works are shown on the flood hazard map as providing protection during design flood conditions.

Design of structures behind levees and floodwalls shall consider the adequacy of drainage of rainfall, runoff, and other waters behind the levees and floodwalls.

New construction adjacent to flood protective works shall not (1) damage, endanger, or otherwise harm the flood protective works, (2) be in conflict with maintenance and repair operations of the entity responsible for the flood protective works, or (3) significantly increase the potential for trapping of rainfall, runoff, and other waters behind the flood protective works.

1.4.3 Assignment of Flood Design Class to Buildings and Structures For the purpose of applying this standard, each building and structure shall be assigned a Flood Design Class based on the risk to human life, health, and welfare associated with damage or failure due to flooding, and by nature of their occupancy or use, according to Table 1-1. Each building or other structure shall be assigned to the highest applicable Flood Design Class or Classes.

1.4.3.1 Multiple Flood Design Classes Where buildings or other structures are divided into portions with independent structural systems, the Flood Design Class assignment for each portion shall be permitted to be determined independently.

Where building systems, such as required egress, HVAC, or electrical power, for a portion with a higher Flood Design Class pass through or depend on other portions of the building or other structure having a lower Flood Design Class, those portions shall be assigned to the higher Flood Design Class.

1.5 BASIC DESIGN AND CONSTRUCTION REQUIREMENTS

1.5.1 General New construction and substantial improvements shall be designed, constructed, connected, and anchored to resist flotation, collapse, or permanent lateral movement resulting from the action of hydrostatic, hydrodynamic, wind, and other loads during design flood, or lesser, conditions in accordance with requirements of this standard if specified, or if not specified in this standard then in accordance with requirements approved by the authority having jurisdiction. Design shall include the loads and load combinations described in Section 1.6.

Design and construction in flood hazard areas shall account for each of the following in accordance with this standard:

1. Elevation of the structure,
2. Foundation types and site-specific geotechnical factors,
3. Resistance of structures to damage up to and during the design flood,
4. Obstructions or enclosures below elevated structures,
5. Structural members and connections required to resist design loads,
6. Use of flood damage-resistant materials,
7. Floodproofing,
8. Utilities,
9. Means of egress, and
10. Adverse impacts to other structures and property.

1.5.2 Elevation Requirements New construction and substantial improvements shall have the lowest floors (including basements) elevated to or above the DFE in conformance with the requirements of the chapter applicable to the specific flood hazard area.

Enclosed areas that are used solely for parking of vehicles, building access, or storage are not the lowest floor and shall be allowed below the DFE, provided the enclosed areas meet the requirements for enclosed areas applicable to the specific Flood Hazard Area. Structures that are used solely for parking of vehicles, building access, or storage shall be allowed below the DFE, provided the structures meet the requirements of Section 9.4.

Nonresidential structures and nonresidential portions of mixed-use structures shall be allowed to have the lowest floor (including basements) below the DFE, provided the structures meet the dry floodproofing requirements in Chapter 6.

1.5.3 Foundation Requirements Foundations of structures shall be designed and constructed to support the structures during design flood conditions and shall provide the required support to prevent flotation, collapse, or permanent lateral movement under the load combinations specified in Section 1.6.2. Any part of the foundation that is below the minimum elevations specified by Table 2-1 or Table 4-1, as applicable, and that provides structural support to the structure shall meet applicable foundation requirements in this standard.

1.5.3.1 Geotechnical Considerations Foundation design shall be based on the geotechnical characteristics of the soils and strata below the structure and on interactions between the soils and strata and the foundation. Foundation design shall account for instability and decreased structural capacity associated with soil consolidation, expansion, or movement; erosion and local scour; liquefaction; and subsidence, as applicable.

Geotechnical information necessary to complete the foundation design shall be obtained through geotechnical investigations of the site or from existing available data, such as investigations conducted at nearby project sites, regional studies conducted by government agencies, or other reliable sources.

1.5.3.2 Foundation Depth The foundation shall extend to a depth based on geotechnical considerations to provide the support described in Sections 1.5.3 and 1.5.3.1, taking into account the erosion and local scour of the supporting soil based on an erosion analysis.

1.5.3.3 Foundation Walls and Wall Footings Foundation walls extending below the minimum elevations specified by Table 2-1 or Table 4-1, as applicable, and foundation wall footings shall be designed and constructed to account for (1) hydrostatic, hydrodynamic, flood-borne debris impact, soil, wind, and other lateral loads acting during design flood conditions, and (2) buoyancy, dead load, live load, and other vertical loads acting during design flood conditions.

Foundation walls, foundation wall footings, and connections between the elevated building and the foundation walls, and between the foundation walls and the foundation wall footings, shall have the strength and stability to resist applied loads and to transfer applied loads to the underlying soils.

1.5.3.4 Piers, Posts, Columns, or Piles Piers, posts, columns, or piles used to elevate a structure above the DFE in flood hazard areas shall comply with all applicable foundation requirements of this standard. In Coastal High Hazard Areas and Coastal A Zones, piers, posts, columns, or piles used to elevate a structure shall be designed and constructed in accordance with Chapter 4. Connections between footings, mat, or raft foundations and piers, posts, and columns shall meet all applicable requirements of this standard.

1.5.4 Use of Fill Fill shall be designed to be stable under conditions of flooding, including rapid rise and rapid drawdown of floodwaters, prolonged inundation, and flood-related erosion and scour. Use of fill in flood hazard areas other than High Risk Flood Hazard Areas shall be in accordance with Section 2.4. Use of fill in Coastal High Hazard Areas and Coastal A Zones shall be in accordance with Section 4.5.4.

1.5.5 Anchorage and Connections The structure, including anchorage and connections, shall be designed to resist effects of

Table 1-1 Flood Design Class of Buildings and Structures

Use or Occupancy of Buildings and Structures	Flood Design Class
Buildings and structures that normally are unoccupied and pose minimal risk to the public or minimal disruption to the community should they be damaged or fail due to flooding. Flood Design Class 1 includes (1) temporary structures that are in place for less than 180 days, (2) accessory storage buildings and minor storage facilities (does not include commercial storage facilities), (3) small structures used for parking of vehicles, and (4) certain agricultural structures.[a]	1
Buildings and structures that pose a moderate risk to the public or moderate disruption to the community should they be damaged or fail due to flooding, except those listed as Flood Design Classes 1, 3, and 4. Flood Design Class 2 includes the vast majority of buildings and structures that are not specifically assigned another Flood Design Class, including most residential, commercial, and industrial buildings.	2
Buildings and structures that pose a high risk to the public or significant disruption to the community should they be damaged, be unable to perform their intended functions after flooding, or fail due to flooding. Flood Design Class 3 includes (1) buildings and structures in which a large number of persons may assemble in one place, such as theaters, lecture halls, concert halls, and religious institutions with large areas used for worship; (2) museums; (3) community centers and other recreational facilities; (4) athletic facilities with seating for spectators; (5) elementary schools, secondary schools, and buildings with college or adult education classrooms; (6) jails, correctional facilities, and detention facilities; (7) healthcare facilities not having surgery or emergency treatment capabilities; (8) care facilities where residents have limited mobility or ability, including nursing homes but not including care facilities for five or fewer persons; (9) preschool and child care facilities not located in one- and two-family dwellings; (10) buildings and structures associated with power generating stations, water and sewage treatment plants, telecommunication facilities, and other utilities which, if their operations were interrupted by a flood, would cause significant disruption in day-to-day life or significant economic losses in a community; and (11) buildings and other structures not included in Flood Design Class 4 (including but not limited to facilities that manufacture, process, handle, store, use, or dispose of such substances as hazardous fuels, hazardous chemicals, hazardous waste, or explosives) containing toxic or explosive substances where the quantity of the material exceeds a threshold quantity established by the authority having jurisdiction and is sufficient to pose a threat to the public if released.[b]	3
Buildings and structures that contain essential facilities and services necessary for emergency response and recovery, or that pose a substantial risk to the community at large in the event of failure, disruption of function, or damage by flooding. Flood Design Class 4 includes (1) hospitals and health care facilities having surgery or emergency treatment facilities; (2) fire, rescue, ambulance, and police stations and emergency vehicle garages; (3) designated emergency shelters; (4) designated emergency preparedness, communication, and operation centers and other facilities required for emergency response; (5) power generating stations and other public utility facilities required in emergencies; (6) critical aviation facilities such as control towers, air traffic control centers, and hangars for aircraft used in emergency response; (7) ancillary structures such as communication towers, electrical substations, fuel or water storage tanks, or other structures necessary to allow continued functioning of a Flood Design Class 4 facility during and after an emergency; and (8) buildings and other structures (including, but not limited to, facilities that manufacture, process, handle, store, use, or dispose of such substances as hazardous fuels, hazardous chemicals, or hazardous waste) containing sufficient quantities of highly toxic substances where the quantity of the material exceeds a threshold quantity established by the authority having jurisdiction and is sufficient to pose a threat to the public if released.[b]	4

[a] Certain agricultural structures may be exempt from some of the provisions of this standard; see Section C1.4.3.

[b] Buildings and other structures containing toxic, highly toxic, or explosive substances shall be eligible for assignment to a lower Flood Design Class if it can be demonstrated to the satisfaction of the authority having jurisdiction by a hazard assessment as described in Section 1.5.3 of *Minimum Design Loads for Buildings and Other Structures* that a release of the substances is commensurate with the risk associated with that Flood Design Class.

vertical loads, including uplift and lateral loads in accordance with the load combinations specified in Section 1.6.2.

Stringers or beams shall be attached to the substructure or directly to piles, columns, piers, and walls with bolted or welded connections such that a continuous load path is maintained.

Washers shall be used under all nuts and bolt heads bearing directly on wood. All nuts, bolts, and washers shall be corrosion resistant. Notches at the tops of timber posts and piles shall not exceed 50% of the cross section of the post or pile.

Adequate anchorage shall be provided for storage tanks, sealed conduits and pipes, lined pits, sumps, and all other similar structures that are subject to flotation or lateral movement during the design flood.

1.6 LOADS IN FLOOD HAZARD AREAS

1.6.1 General Design of structures within flood hazard areas shall be governed by the loading provisions of ASCE 7 *Minimum Design Loads for Buildings and Other Structures* (ASCE/SEI 2010).

Design and construction of structures located in flood hazard areas shall consider all flood-related loads and conditions, including the following: hydrostatic loads, hydrodynamic loads, wave action; debris impact; rapid rise and rapid drawdown of floodwaters; prolonged inundation; alluvial fan flooding; wave-induced and flood-related erosion and local scour; deposition of sediments; ice flows and ice jams; and mudslides in accordance with requirements of this standard if specified, or if not specified in this standard then in accordance with requirements approved by the authority having jurisdiction. Design considerations shall be documented and shall take into account the applicable flood-related loads and conditions, and load combinations that will act on the foundation and the structure.

1.6.2 Combination of Loads Flood loads shall be combined with other loads as specified in ASCE 7 *Minimum Design Loads for Buildings and Other Structures* (ASCE/SEI 2010), either by using the allowable stress design method load combinations or by using the strength design method load combinations.

CHAPTER 2

BASIC REQUIREMENTS FOR FLOOD HAZARD AREAS THAT ARE NOT IDENTIFIED AS COASTAL HIGH HAZARD AREAS AND COASTAL A ZONES

2.1 SCOPE

The requirements of this section shall apply to new construction and substantial improvements in Flood Hazard Areas, including High Risk Flood Hazard Areas that are not identified as Coastal High Hazard Areas and Coastal A Zones.

2.2 DEVELOPMENT IN FLOODWAYS

Structures and fill shall not be constructed or placed in floodways unless it is demonstrated that those structures and fill will not (1) increase the flood level during occurrence of the base flood discharge, and (2) reduce the conveyance of the floodway.

If the design flood elevation has been determined and a floodway has not been designated, structures and fill shall not be constructed or placed unless it has been demonstrated that the cumulative effect of proposed structures and fill, combined with all other existing and anticipated development, will not increase the base flood elevation more than 1 ft.

2.3 ELEVATION REQUIREMENTS

Structures shall have the lowest floor (including basements) elevated in conformance with the minimum elevation requirements of Table 2-1. Enclosed areas used solely for parking of vehicles, building access, or storage and that comply with Section 2.7 are allowed below elevated buildings. Elevation requirements for other building components are found in Chapters 5, 6, and 7.

Exception: Nonresidential structures with the lowest floor (including basement) below the minimum elevation specified in Table 2-1 and nonresidential portions of mixed-use structures with the lowest floor (including basement) below the minimum elevation specified by Table 2-1 shall be allowed in conformance with the dry floodproofing requirements of Section 6.2.

2.4 USE OF FILL

Fill shall not be placed in floodways unless in compliance with the requirements of Section 2.2.

2.4.1 Structural Fill Structural fill shall not be used unless design and construction of the structural fill accounts for (1) consolidation of the underlying soil under the weight of the fill and the structure, (2) differential settlement due to variations in fill composition and characteristics, and (3) slope stability and erosion control during conditions of the design flood.

Fill used for structural support or protection shall be suitable for its intended use. Structural fill used to support or protect a structure shall be placed in lifts of not more than 12-in. loose thickness, with each lift compacted to at least 95% of its maximum standard proctor density (ASTM 2012f) or 90% of its maximum modified proctor density (ASTM 2012e), unless otherwise required by the building code or specified in a geotechnical investigation report or a soils engineering report prepared by a qualified registered design professional and approved by the authority having jurisdiction.

The side slopes of structural fill shall be no steeper than 1 on 1.5 (vertical/horizontal). Structural fill, including side slopes, shall be protected from erosion under flood conditions up to and including the design flood.

2.5 SLABS-ON-GRADE AND FOOTINGS

2.5.1 Use of Slabs-on-Grade Use of slabs on grade is acceptable if the slabs are constructed on structural fill that is placed in conformance with Section 2.4 or constructed on undisturbed soil of adequate bearing capacity. The top of the slab shall be at or above the elevation specified in Table 2-1. If turned down to act as footings, the bottom of the turned-down edges of the slab shall be installed at or below the depth of expected local scour and erosion.

Reinforcement shall be sufficient to prevent breakup of the slab during design flood conditions, even if the soil under the slab is undermined by erosion. Slabs-on-grade constructed on structural fill shall be placed so that there is no loss of supporting soil during the design flood conditions.

Slabs that conform to Section 9.3 shall be permitted.

Exception: Slabs located under elevated buildings and used as parking slabs or floors of enclosed areas that comply with Section 2.7 shall be permitted.

2.5.2 Footing Design Footings shall support the structure during design flood conditions, including prolonged inundation and local scour and erosion if expected during design flood conditions to prevent flotation, collapse, and lateral movement.

Footings that are intended also to act as grade beams shall comply with the provisions of Section 4.5.10.

2.6 FOUNDATION WALLS

Foundation walls that form enclosed areas shall be designed to meet the requirements of this section.

Masonry walls shall be designed and constructed in accordance with the requirements of ACI 530/ASCE 5/TMS 402, *Building Code Requirements and Specifications for Masonry Structures and Related Commentaries* (ACI 2013). The empirical design of masonry contained in Chapter 5 of ACI 530/ASCE 5/TMS 402, *Building Code Requirements and Specifications for Masonry Structures and Related Commentaries* (ACI 2013) is not permitted to be used.

Concrete shall be designed and constructed in accordance with the requirements of ACI 318, *Building Code Requirements for Structural Concrete and Commentary* (ACI 2014).

Table 2-1 Minimum Elevation of the Top of Lowest Floor—Flood Hazard Areas Other Than Coastal High Hazard Areas,[a] Coastal A Zones,[a] and High Risk Flood Hazard Areas[a]

Flood Design Class[b]	Minimum Elevation, Relative to Base Flood Elevation (BFE) or Design Flood Elevation (DFE)
1[c]	DFE
2[d]	BFE + 1 ft or DFE, whichever is higher
3[d]	BFE + 1 ft or DFE, whichever is higher
4[d]	BFE + 2 ft or DFE, or 500-year flood elevation, whichever is higher

[a]Minimum elevations shown in Table 2-1 do not apply to Coastal High Hazard Areas and Coastal A Zones (see Table 4-1). Minimum elevations shown in Table 2-1 apply to other high risk flood hazard areas unless specific elevation requirements are given in Chapter 3 of this standard.
[b]See Table 1-1 for Flood Design Class descriptions.
[c]Flood Design Class 1 structures shall be allowed below the minimum elevation if the structure meets the wet floodproofing requirements of Section 6.3.
[d]For nonresidential buildings and nonresidential portions of mixed-use buildings, the lowest floor shall be allowed below the minimum elevation if the structure meets the dry floodproofing requirements of Section 6.2.

Table 2-2 Flood Opening Coefficient of Discharge[a]

Opening Shape and Condition	*c*
All shapes, partially obstructed during design flood[b]	0.20
Circular, unobstructed during design flood	0.60
Rectangular, long axis horizontal, short axis vertical, unobstructed during design flood	0.40[c]
Square, unobstructed during design flood	0.35
Rectangular, short axis horizontal, long axis vertical, unobstructed during design flood	0.25[d]
Other shapes, unobstructed during design flood	0.30

[a]Different coefficients of discharge shall be permitted: (1) where a designer has performed detailed, opening-specific calculations, a coefficient of discharge up to 10% different than given in Table 2-2 shall be permitted; or (2) where laboratory testing or numerical modeling of flow through the opening has been conducted, the resulting coefficient of discharge shall be permitted. In no case shall a coefficient of discharge >0.60 be permitted.
[b]Openings shall be classified as partially obstructed if louvers, blades, screens, grilles, faceplates, or other covers or devices are present during the design flood.
[c]When the horizontal dimension is twice or more the vertical dimension, use 0.4; as the dimensions approach a square, interpolate from 0.4 to 0.35.
[d]When the horizontal dimension is half or less the vertical dimension, use 0.25; as the dimensions approach a square, interpolate from 0.25 to 0.35.

2.7 ENCLOSURES BELOW THE DESIGN FLOOD ELEVATION

Enclosed areas that are used solely for parking of vehicles, building access, or storage shall be permitted below the DFE, provided the enclosed areas meet the requirements of this section.

2.7.1 Required Openings in Foundation Walls and Walls of Enclosures Foundation walls and exterior walls that form enclosures below the DFE that do not meet the dry-floodproofing requirements of Section 6.2 shall contain openings to allow for automatic entry and exit of floodwaters during design flood conditions. Openings shall meet the requirements of Section 2.7.2 and Section 2.7.3.

2.7.1.1 Openings in Breakaway Walls Openings to allow for the automatic entry and exit of floodwaters during design flood conditions shall be installed in breakaway walls in all flood hazard areas. Openings shall meet the requirements of Section 2.7.2 and be installed in accordance with Section 2.7.3.

2.7.2 Design of Openings Openings shall meet the non-engineered opening requirements of Section 2.7.2.1 or the engineered opening requirements of Section 2.7.2.2. Installation of all openings shall meet the requirements of Section 2.7.3.

2.7.2.1 Non-Engineered Openings Non-engineered openings shall meet the following criteria: (1) The total net open area of all openings shall be at least 1 sq in. for each sq ft of enclosed area, where the enclosed area is measured on the exterior of the enclosure walls; (2) openings shall not be less than 3 in. in any direction in the plane of the wall; and (3) the presence of louvers, blades, screens, and faceplates or other covers and devices shall not block or impede the automatic flow of floodwaters into and out of the enclosed areas and shall be accounted for in the determination of the net open area.

2.7.2.2 Engineered Openings Engineered openings shall meet the following criteria:

1. Each individual opening shall be designed to allow automatic entry and exit of floodwaters during design flood or lesser flood conditions;
2. The performance of engineering openings shall account for the presence of louvers, blades, screens, grilles, faceplates, or other covers and devices;
3. Openings shall not be less than 3 in. in any direction in the plane of the wall;
4. The performance of engineered openings shall ensure that the difference between the exterior and interior floodwater levels shall not exceed 1 ft;
5. In the absence of reliable data on the rates of rise and fall, assume a minimum rate of rise and fall of 5 ft/h; where an analysis indicates the rates of rise and fall are greater than 5 ft/h, the total net area of the required openings shall be increased to account for the higher rates of rise and fall; where an analysis indicates the rates of rise and fall are less than 5 ft/h, the total net area of the required openings shall remain the same or shall be decreased to account for the lower rates of rise and fall; and
6. The minimum total net area of the required openings in enclosure walls shall be calculated using the equation

$$A_o = 0.033\,(1/c)(R)(A_e)$$

where

A_o = the total net area of openings required (in.^{2})
0.033 = coefficient (in.$^2 \cdot$ h/ft^3) corresponding to a factor of safety of 5.0
c = opening coefficient of discharge given in Table 2-2
R = worst case rate of rise and fall (ft/h)
A_e = the total enclosed area (ft^2).

2.7.3 Installation of Openings Installation of openings shall meet the following criteria:

1. Each enclosed area shall have a minimum of two openings,
2. Openings shall be in at least two walls of each enclosed area,
3. The bottom of each opening shall be no more than 1 ft above the higher of the final interior grade or floor and the finished exterior grade immediately under each opening, and
4. Openings meeting requirements of Section 2.7.2.1 or Section 2.7.2.2 that are installed in doors and windows are permitted.

CHAPTER 3
HIGH RISK FLOOD HAZARD AREAS

3.1 SCOPE

The requirements of Chapter 3 shall apply to new construction and substantial improvements in High Risk Flood Hazard Areas subject to one or more of the following hazards: alluvial fan flooding, flash floods, mudslides, erosion, high-velocity flows, high-velocity wave action, breaking wave heights greater than or equal to 1.5 ft (Coastal High Hazard Area and Coastal A Zone) and damage-causing ice or debris. In addition to the requirements of Chapter 3, the basic requirements of Chapter 2 shall apply to High Risk Flood Hazard Areas other than Coastal High Hazard Areas and Coastal A Zones.

3.2 ALLUVIAL FAN AREAS

New construction and substantial improvements shall not be constructed at the apex of an alluvial fan, in the fan's meandering flow paths, or in areas of the fan that have characteristics and evidence that the natural processes that form alluvial fans are active on the fan's surface, including braided channels, erratic flow paths, and sediment transport. Construction in other areas of the alluvial fan shall meet the following requirements:

1. The elevation of the lowest floor shall be a minimum of 1 ft above the highest adjacent grade, or higher, if required on a community's flood hazard map;
2. Foundations shall be designed and constructed to resist scour caused by the actual flow velocity but not less than 5 ft/s. Determination of actual flow velocities shall be based on a review of a community's flood hazard map and flood hazard study or on hydraulic calculations; and
3. Design and construction shall resist all load combinations specified in Section 1.6.2.

3.2.1 Protective Works in Active Alluvial Fan Areas New construction and substantial improvements shall not be allowed in active alluvial fan areas that have characteristics and evidence that the natural processes that form alluvial fans are active on the fan's surface unless protective works (whole alluvial fan flood damage reduction project) have been designed and constructed (1) to safely pass the design flood at the apex, within the capacity of the constructed channel(s), (2) such that it does not divert flood flows and debris toward other structures nor increase flood velocities and depths elsewhere on the alluvial fan, and (3) such construction satisfies the requirements of Section 1.4.2, and a maintenance and operations plan for the protective works is provided.

3.3 FLASH FLOOD AREAS

New construction and substantial improvements shall not be constructed in areas subject to flash flooding equal to or less than design flood conditions.

Areas suspected of being subject to flash floods shall be investigated to obtain historical information on past events. The investigation shall also include analysis of historic rainfall and runoff data for the watershed. Results of such analyses shall be documented in an engineering report, which defines the methodology and data used to conclude whether the area in question has the potential for flash flooding.

3.3.1 Protective Works in Flash Flood Areas New construction and substantial improvements shall not be constructed in areas subject to flash floods unless protective works have been determined to provide protection during the design flood event, where such construction satisfies the requirements of Section 1.4.2, and where a maintenance and operations plan for the protective works has been provided.

3.4 MUDSLIDE AREAS

New construction and substantial improvements shall not be constructed in areas subject to mudslides during periods of rainfall and runoff. Areas suspected of being subject to inundation by mudslides shall be investigated to obtain historical information on past flood events. The investigation also shall include analysis of the source area for potential overland or channel erosion, bank failure, hillslope failure, and rainfall/runoff potential. Results of such analyses shall be documented in an engineering report, which defines the methodology and data used to conclude whether the area in question has potential for future mudslides.

3.4.1 Protective Works in Mudslide Areas New construction and substantial improvements shall not be constructed in areas subject to mudslides unless protective works have been determined to provide protection during the design flood event, such construction satisfies the requirements of Section 1.4.2, and a maintenance and operations plan for the protective works has been provided.

3.5 EROSION-PRONE AREAS

New construction and substantial improvements shall not be constructed within flood hazard areas subject to erosion from such phenomena as caving banks, meandering streams, or eroding shorelines, where such erosion is predicted to affect the structure unless the structure is protected as specified in Section 3.5.1.

Erosion-prone areas shall be determined by analyzing available studies, historical data, watershed trends, average annual erosion rates, wave effects, flood velocities and duration of flow, geotechnical data, and existing protective works. Results of these analyses shall be documented in an engineering report, which defines the data and methodology used to identify erosion-prone areas.

3.5.1 Protective Works in Erosion-Prone Areas The limits of an erosion-prone area shall be subject to revision where protective works have been designed and constructed to control erosion processes during all flow and wave conditions up to and including the design flood, and where a maintenance and operations plan for the protective works has been provided.

3.6 HIGH VELOCITY FLOW AREAS

High velocity flow areas shall be identified from a community's flood hazard map, flood hazard study, or from hydraulic analyses. The results of such analyses shall be documented in an engineering report, which defines the methodology and data used to conclude whether a site is susceptible to high velocity flows.

3.6.1 Protective Works in High Velocity Flow Areas New construction and substantial improvements shall not be constructed in high velocity flow areas unless protective works have been determined to provide protection during the design flood event, such construction satisfies the requirements of Section 1.4.2, and a maintenance and operations plan for the protective works has been provided.

3.7 AREAS SUBJECT TO WAVE ACTION

3.7.1 Coastal High Hazard Areas and Coastal A Zones New construction and substantial improvements shall not be constructed in Coastal High Hazard Areas and Coastal A Zones unless the design and construction meet the requirements of Chapter 4.

3.7.2 Other High Velocity Wave Action Areas Other high velocity wave action areas include noncoastal flood hazard areas subject to wind-driven waves greater than or equal to 3 ft in height, including large lakes and expansive riverine flood hazard areas. New construction and substantial improvements shall not be constructed in other areas susceptible to high velocity wave action unless the design and construction meet the requirements of Chapter 4.

3.8 ICE JAM AND DEBRIS AREAS

New construction and substantial improvements shall not be constructed within flood hazard areas that are subject to transportation of damage-causing ice or debris during floods up to and including the design flood.

The potential for ice or debris capable of inducing or causing loads exceeding design loads shall be identified from a community's flood hazard map or flood hazard study or from hydraulic and other analyses. The results of such analyses shall be documented in an engineering report, which defines the methodology and data used to conclude whether a site is susceptible to ice jams and debris effects.

3.8.1 Protective Works in Ice Jam and Debris Areas New construction and substantial improvements in ice jam and debris areas shall have protective works to provide protection during the design flood event and meet the requirements of Section 1.4.2. The maintenance and operations plan for the protective works shall be provided.

CHAPTER 4
COASTAL HIGH HAZARD AREAS AND COASTAL A ZONES

4.1 SCOPE

The requirements of Chapter 4 shall apply to new construction and substantial improvements in Coastal High Hazard Areas and Coastal A Zones.

4.1.1 Identification of Coastal High Hazard Areas and Coastal A Zones For the purposes of this standard, "Coastal High Hazard Areas" shall mean those locations (1) where an area has been designated as subject to high velocity wave action on a community's flood hazard map (V Zones), (2) where the still-water depth of the base flood above the eroded ground elevation is greater than or equal to 3.8 ft (i.e., sufficient to support a wave height equal to or greater than 3 ft and where conditions are conducive to the formation and propagation of such waves), or (3) where the eroded ground elevation under base flood conditions is 3 ft or more below the maximum wave runup elevation.

For the purposes of this standard, "Coastal A Zones" shall mean those locations (1) landward of the V Zone or shoreline and seaward of the Limit of Moderate Wave Action (LiMWA), if the LiMWA is delineated on a FIRM, or (2) designated by the authority having jurisdiction.

4.2 GENERAL

Designs for Coastal High Hazard Areas and Coastal A Zones shall account for local scour and erosion and shall be designed to resist loads from the following:

1. Waves breaking against the bracing, side of the structure, and underside of the structure,
2. Drag, inertia, and other wave-induced forces acting on structural members supporting elevated structures,
3. Uplift forces from breaking waves striking the undersides of structures, and
4. Wave runup forces including those deflected by the structure.

4.3 SITING

Within Coastal High Hazard Areas and Coastal A Zones:

1. New construction, not including substantial improvements, shall be located landward of the reach of mean high tide;
2. New construction and substantial improvements shall be sited landward of shoreline construction setbacks, where applicable; and
3. New construction and substantial improvements shall not remove or otherwise alter sand dunes and mangrove stands, unless an engineering report documents that the alterations will not increase potential flood damage by reducing the wave and flow dissipation characteristics of the sand dunes or mangrove stands.

4.4 ELEVATION REQUIREMENTS

The bottom of the lowest horizontal structural member of the lowest floor shall be elevated in conformance with the minimum requirements of Table 4-1. The actual required height above the DFE shall be determined by the Flood Design Class of the structure. Piles, pile caps, footings, mat or raft foundations, grade beams, columns, bracing, and shear walls designed and constructed in accordance with Section 4.5 shall not be required to meet the elevation requirements of Table 4-1.

4.5 FOUNDATION REQUIREMENTS

4.5.1 General Foundation systems located in Coastal High Hazard Areas and Coastal A Zones shall be designed to minimize forces acting on the foundations, to minimize damage to the foundations and the elevated structures, and to adequately transfer all loads specified by Section 1.6 and imposed on the foundations and elevated structures to the supporting soils. Foundation systems shall be free of obstructions and attachments that will transfer flood forces to the structural system or that will restrict or eliminate free passage of high velocity flood waters and waves during design flood conditions unless provided for in Section 4.5.11 or Section 4.5.12.

Unless provided for in Section 4.5.1.1, foundations in Coastal High Hazard Areas constructed on erodible soils shall consist of piles, drilled shafts, caissons, or other deep foundations.

Unless provided for in Section 4.5.1.2, foundations in Coastal A Zones constructed on erodible soils shall consist of piles, drilled shafts, caissons, or other deep foundations.

Foundation systems shall extend upward to elevate structures as required by Table 4-1.

Columns shall be connected to and extend upward from footings, mats, rafts, or concrete slabs to elevate structures as required by Table 4-1, provided the footings, mats, rafts, or concrete slabs meet the requirements of Section 4.5.8.

Shear walls shall comply with the requirements of Section 4.5.12.

Stem walls shall comply with the requirements of Section 4.5.13.

4.5.1.1 Shallow Foundations in Coastal High Hazard Areas In Coastal High Hazard Areas where surface or subsurface conditions consist of nonerodible soils or rock that prevent deep foundations, shallow foundations including spread footing, mat and raft foundations shall be permitted, provided that the foundations (1) meet the requirements of Section 4.5.8; (2) are

Table 4-1 Minimum Elevation of Bottom of Lowest Supporting Horizontal Structural Member of Lowest Floor—Coastal High Hazard Areas and Coastal A Zones

Flood Design Class[a]	Minimum Elevation, Relative to Base Flood Elevation (BFE) or Design Flood Elevation (DFE)
1	DFE
2	BFE + 1 ft or DFE, whichever is higher
3	BFE + 2 ft or DFE, whichever is higher
4	BFE + 2 ft or DFE, or 500-year flood elevation, whichever is higher

[a] See Table 1-1 for Flood Design Class descriptions.

anchored, if necessary to prevent sliding, uplift, or overturning; and (3) have sufficient strength to withstand forces from the combination of loads in Section 1.6.2.

4.5.1.2 Shallow Foundations in Coastal A Zones In Coastal A Zones, shallow foundations including spread footing, mat and raft foundations shall be permitted, provided the foundations (1) meet the requirements of Section 4.5.8; and (2) will prevent sliding, uplift, or overturning when exposed to the combination of loads in Section 1.6.2.

4.5.2 Special Geotechnical Considerations In addition to the requirements of Section 1.5.3, foundation design shall account for instability and decreased structural capacity associated with saturated soils and with erosion because of wind, waves, currents, local scour, storm-induced erosion, and shoreline movement.

4.5.3 Foundation Depth The foundation shall extend to a depth sufficient to provide the support required in Section 1.5.3, taking into account the erosion and scour of the supporting soil during the design flood and shoreline movement, as predicted by an erosion analysis.

4.5.4 Use of Fill Fill material used for structural support shall not be permitted in Coastal High Hazard Areas and Coastal A Zones. Placement of nonstructural fill for minimal site grading and landscaping and to meet local drainage requirements shall be permitted. Placement of nonstructural fill under and around a structure for dune construction or reconstruction shall be permitted if an engineering report documents that the fill will not result in wave runup, ramping, or deflection of floodwaters that cause damage to structures.

4.5.5 Deep Foundations Except as provided for under Section 4.5.1, all foundations constructed in erodible soils shall be founded on piles, drilled shafts, caissons, or other types of deep foundations. Piles that are jetted or installed in an augured excavation shall be seated by driving.

All foundations shall resist lateral and vertical loads specified by Section 1.6. Soils susceptible to erosion and local scour shall not be considered in determining the load resistance of the foundation. Foundations shall extend to a depth at least 10 ft below mean water level (MWL) unless the design demonstrates that pile penetration to a shallower depth will provide the support and stability required by Section 4.5.3 while resisting loads specified by Section 1.6. In the event that unexpected conditions are encountered during construction and refusal or design friction capacity is not reached during pile installation, additional geotechnical investigations and a revised pile design shall be completed.

The design shall consider that local scour, erosion, and liquefaction of the erodible soil during design flood conditions will render as nonsupportive the soil at least to a depth of two times the pile diameter (round pile) or two times the diagonal (rectangular pile) below the eroded ground elevation at the point of pile penetration after flood- or storm-induced erosion has been considered. This nonsupportive soil shall not be considered in the design. Calculation of local scour effects during design flood conditions shall include the interactive effects of pilings or other foundation elements in close proximity to one another.

4.5.5.1 Attachments to Piles Pile design shall account for the additional loads resulting from any attachments to the piles and shall account for increased local scour and erosion around the base of the pile.

Bracing used for lateral support of pile foundation systems shall be in accordance with Section 4.5.11.

4.5.5.2 Piles Terminating in Pile Caps or Grade Beams That are At or Below Grade Foundations composed of a number of single piles or pile clusters terminating in reinforced concrete pile caps or grade beams at or below grade shall be designed for the combination of loads in Section 1.6.2. Embedment of the pile into the pile cap or embedment of the pile into the grade beam shall be sufficient to resist separation of the two. Embedment shall not be less than that specified by Section 4.5.6.

The pile cap or the grade beams shall be designed and constructed to be structurally stable without relying on supporting soil around or under the caps or under the grade beams. All connections between piles, pile caps, grade beams, and column connections shall be designed to withstand the imposed loads from the elevated structure and the expected hydrodynamic forces including wave and debris impact.

4.5.5.3 Piles Extending to Superstructure (Structure Framing) The portion of a pile that extends above the eroded ground elevation to elevate a structure shall be designed as a column. Pile spacing shall take into account the design bearing capacity, uplift, and overturning resistance.

Bracing shall be provided in accordance with Section 4.5.11.

4.5.5.4 Wood Piles Wood piles shall be preservative treated in accordance with Chapter 5. Consideration shall be given to the use of pile fittings at the butt and tip and at designated intervals along the pile length for the protection of piles during installation. Round timber piles shall conform to ASTM D25 *Standard Specification for Round Timber Piles* (ASTM 2012c).

Round piles shall have a minimum diameter of 8 in. at the tip. The minimum size for square piles shall be 8 in. on a side where the exposed pile length (after accounting for scour and erosion) is equal to or less than 12 ft and shall be 10 in. on a side where the exposed pile length (after accounting for local scour and erosion) is greater than 12 ft.

Where wood piles are connected directly to beams supporting an elevated lowest floor, the connection of the beam to the pile shall consist of at least two bolts sized to resist the forces resulting from the combinations of loads in Section 1.6.2. Bolts, nuts, and washers used to secure beams to piles shall be hot-dip galvanized steel or stainless steel in accordance with Section 5.2.1.

4.5.5.5 Steel H Piles Steel piles of rolled HP sections or builtup sections shall consist of a corrosion resistant material or be protected from corrosion by a corrosion resistant coating or by cathodic protection in accordance with Chapter 5. Builtup sections shall have a web thickness equal to the flange thickness; the web and flange shall be continuously welded together.

The metal thickness shall be based on the loss of section because of corrosion, unless corrosion protection is provided in the form of concrete, bituminous or plastic (epoxy) coatings, or cathodic protection. The minimum allowable thickness of the metal shall be 0.4 in. Damage to coatings during installation shall be avoided, and, if damage occurs, repairs shall be made in accordance with manufacturers' recommendations and applicable codes, standards, and regulations.

Pile tip reinforcing, splicing, fittings, and cap plates shall be provided, as required.

4.5.5.6 Concrete-Filled Steel Pipe Piles and Shells Steel components of concrete-filled steel pipe piles and shells shall be protected with a corrosion resistant coating in accordance with Chapter 5.

Pipe for concrete-filled steel pipe piles shall conform to ASTM A252 *Standard Specification for Welded and Seamless Steel Pipe Piles* (ASTM 2010b). Pile tip reinforcing, splicing, fittings, and cap plates shall be specified as required. Flat steel plates closing the tip of pipe piles shall be of a diameter not greater than 3/4 in. more than the outside diameter of the pipe.

The slump of concrete and maximum size of coarse aggregate used in concrete to be placed in piles shall be specified with consideration given to the difficulty of placement conditions.

4.5.5.7 Prestressed Concrete Piles and Precast Concrete Piles Prestressed concrete piles and precast concrete piles shall be designed, manufactured, and installed in accordance with Chapter 5.

Pile dimensions shall be 10 in. minimum for piles of uniform section and 8 in. minimum for tapered piles.

For piles subject to exposure from brackish water, seawater, or spray from these sources, cover for reinforcement shall be not less than 3 in. for precast concrete piles and not less than 2-1/2 in. for prestressed concrete piles.

4.5.5.8 Cast-in-Place Concrete Piles Concrete and steel reinforcement used in cast-in-place concrete piles shall conform to the requirements of ACI 318 *Building Code Requirements for Structural Concrete and Commentary* (ACI 2014).

4.5.6 Pile Design

4.5.6.1 Pile Capacity Piles shall be designed to carry the loads imposed by the combinations of loads in Section 1.6.2 and to withstand installation forces.

The minimum flexural resistance of the pile shall not be less than the design axial load on the pile times an eccentricity of 10% of the equivalent pile diameter.

The minimum required lateral resistance of an individual pile shall be at least 5% of the axial load on the pile.

4.5.6.2 Capacity of the Supporting Soils Soil values pertaining to friction, end bearing resistance, and settlement of single piles and pile groups shall be based on the geotechnical characteristics of the soil as required by Section 1.5.3.1.

For piles spaced more than three pile diameters apart, measured center to center, the diameter of the soil that shall be assumed reacting laterally on each pile shall have a maximum equivalent diameter equal to three times the diameter of the pile.

4.5.6.3 Minimum Penetration Pile penetration into acceptable bearing strata shall be a minimum depth sufficient to allow distribution of the pile load to the supporting soils, including a consideration for reduction in soils because of the effects of local scour and erosion in accordance with Section 4.5.3.

4.5.6.4 Foundation Pile Spacing Pile spacing shall be not less than 8 ft as measured center to center unless otherwise justified by a geotechnical analysis and the foundation design.

4.5.6.5 Wood Pile Connections For wood piles, bolts for cap-to-pile or beam-to-pile connections shall be 5/8 in. in diameter minimum. Bolts shall be staggered with a maximum bolt hole diameter of 1/16 in. greater than the bolt diameter. The dimension from the edge of the holes to the pile or beam edge shall be 2 in. minimum. Notching of pile tops shall not exceed 50% of the pile cross section. Other pile-to-beam fasteners or connectors are acceptable, provided they are demonstrated to be equivalent to the bolt. Pile-to-beam attachment shall not depend on nailed connections.

For wood piles not in tension and connected to wood caps, the tops of the piles shall be secured to the caps with spiral-drive drift bolts, metal plates, or bolted wood scabs.

For wood piles not in tension and connected to concrete caps, the tops of the piles shall have a minimum of 4 in. embedment into the concrete pile caps.

For wood piles in tension, piles shall be embedded into pile caps a minimum of 6 in., with a connection designed for tension made to pile caps. Connections to wood caps shall be made with wood scabs or metal straps and headed bolts.

Connections of wood piles in tension to concrete pile caps shall have the tops of the piles embedded to satisfy requirements of shear stress in the wood and shear stress in the concrete. Connections shall be made with metal straps, headed bolts, or other forms of positive tension resisting devices that develop the necessary shear in the concrete without causing failure of the wood.

4.5.6.6 Steel Pile Connections For steel piles not in tension and where structural design of piles depends on bending in the piles for stability, the tops of steel piles shall be tied into concrete caps with reinforcing steel or structural sections welded to the pile and lapped with the cap reinforcement. The minimum lap shall be 6 in.

For steel piles in tension, the tops of steel piles shall be tied into concrete caps with reinforcing steel or structural sections welded to the pile and lapped with the cap reinforcement. The minimum lap shall be 6 in. Bond stress between concrete and steel shall not exceed $0.02 f'_c$, where f'_c is the specified compressive strength of concrete.

4.5.6.7 Concrete Pile Connections For concrete piles not in tension, the tops of the piles shall have a minimum 3 in. embedment into the pile cap.

For concrete piles in tension, the tops of concrete piles in concrete caps shall be adequately doweled and embedded into the cap to resist tension loads. The tops of piles shall have a minimum 3 in. embedment into the pile cap.

4.5.6.8 Pile Splicing Splices shall be constructed to provide and maintain the alignment and position of the component parts of the pile during installation and subsequent thereto. Splices shall be of adequate strength to transmit the axial and lateral loads, and the moments at the section involved.

Pile splices that cannot be inspected visually after pile installation shall develop the greater of at least 50% of the capacity of the pile or the moment, shear, and tension that would result from an assumed eccentricity of the resultant pile load of 3 in.

In areas subject to high velocity wave action and subject to flood-related erosion, wood pile splices shall be made below the scour elevation, as determined by an erosion analysis.

4.5.6.9 Mixed Types of Piling and Multiple Types of Installation Methodology The use of mixed types or capacities of piling and different types of installation equipment or methods shall consider, in addition to the relative lateral load capacities, an analysis of the additional effects on the superstructure of differential elastic shortening and settlement.

4.5.7 Posts, Piers, and Columns Columns, including wood posts, reinforced masonry columns and piers, and reinforced concrete columns and piers, as well as associated connections, shall be designed and constructed to resist wind, water, wave, erosion, and other flood-related forces in accordance with Section 1.6. Column spacing shall take into account the design bearing capacity, uplift, and overturning resistance but shall be spaced not less than 8 ft center to center. Where founded on erodible soils, supports shall extend to a depth at least 10 ft below mean water level or shall be supported by and anchored to a pile, spread footing, mat, or raft foundation meeting the requirements of Section 4.5.1.

4.5.7.1 Wood Posts Wood posts shall be 8 in. × 8 in. minimum for square posts or 10 in. in diameter for round posts. Wood posts shall be preservative treated in conformance with Chapter 5. Where founded on nonerodible soil, posts shall be adequately embedded and anchored to the footing, which shall be anchored to the substrata to prevent pullout during design flood conditions.

4.5.7.2 Reinforced Masonry Columns Reinforced masonry columns shall comply with ACI 530/ASCE 5/TMS 402, *Building Code Requirements and Specifications for Masonry Structures and Related Commentaries* (ACI 2013).

4.5.7.3 Reinforced Concrete Columns Concrete columns shall comply with ACI 318, *Building Code Requirements for Structural Concrete and Commentary* (ACI 2014).

4.5.8 Footings, Mats, Rafts, and Concrete Slabs That Support Columns or Walls Footings, mats, rafts, and concrete slabs that support columns or walls shall be permitted if they conform to the requirements of Section 4.5.1. Foundation elements shall be reinforced, shall be supported directly on undisturbed soil, and the top of the element shall be below the eroded ground elevation. The design shall provide for the effects of local scour and erosion, forces applied to the foundation from the elevated structure, and lateral and vertical loads from flood forces applied to the foundation elements.

4.5.9 Pile Caps The tops of pile caps shall be at or below grade. The design shall provide for the effects of local scour and erosion. Piles shall be designed to carry the total superimposed vertical load from the structure above and from the pile caps with no allowance made for the supporting value of the soil under the pile caps. Pile caps and piles shall be designed to resist lateral flood loads acting on pile caps and pile sections exposed by erosion and local scour.

4.5.10 Grade Beams Grade beams shall be constructed with their upper surface at or below natural grade, shall be structurally connected to the primary foundation system to provide additional lateral support, and shall be structurally independent of decks, patios, and concrete pads. Grade beams shall be designed to perform their structural function without the aid of supporting earth and during exposure to flood forces.

Design of grade beams and other foundation elements shall provide for the effects of lateral bracing offered by grade beams on the structure.

Grade beams shall be attached firmly to vertical members to transfer all vertical and lateral forces acting on the grade beam as a result of flooding, including wave action and the effects of debris, local scour, and erosion. The design of the grade beam members shall include biaxial beam action to support the vertical and lateral forces acting the full length of members in the event they are undermined by local scour and erosion.

4.5.11 Bracing Unless the design determines bracing is not required, bracing shall be provided where more than 8 ft of the pile or column is exposed above the eroded ground elevation. Bracing and its connections to the primary vertical structural members shall be designed to withstand the lateral force of waves and debris impact. The braces shall be designed to resist forces acting both in the plane of brace resistance and perpendicular to the plane of brace resistance. The foundation system shall be designed to account for the additional loads transferred from the bracing to the main supporting structures. Cross bracing perpendicular to the primary direction of wave and hydrodynamic forces shall be restricted to tensile bracing using steel rods or steel cables. Cross bracing parallel to the primary direction of wave and hydrodynamic forces shall not be restricted.

Steel rods used for cross bracing shall be galvanized or of noncorrosive material with a minimum diameter of 1/2 in. An equivalent system of corrosion resistant high tensile steel cables and turnbuckles may be used in lieu of solid rods. Where rods or cables are used for bracing, they shall be tied together with a clamp at the center cross-point.

The smallest nominal dimension of any wood member used for compression cross bracing shall be not less than 4 in. The smallest nominal dimension of any wood member used for tension-only cross bracing shall be 2 in. Wood bracing including knee braces shall be preservative treated in conformance with Chapter 5.

Cross bracing shall be attached to the main vertical structural elements with bolts, nuts, and plate washers. Connections to the main vertical structural elements shall be within 12 in. of the lowest elevated floor support beams and as near to grade as possible.

Knee braces are permitted in all directions relative to flood forces and shall extend not more than 3 ft below the DFE.

4.5.12 Shear Walls Shear walls below the DFE shall be oriented parallel to the direction of wave approach where possible and shall be staggered so as not to form a continuous shear wall or an enclosed area. For the purpose of this provision, parallel shall mean less than or equal to ±20° from the direction of wave approach. If shear walls are oriented other than parallel to the direction of wave approach, an unobstructed area equal to 1/2 the area blocked by the shear wall shall be provided adjacent to each shear wall.

Openings are permitted in shear walls to relieve flood forces. The size and location of the openings are to be determined by engineering analysis and if openings are specified, the design of the shear wall shall account for the openings.

4.5.13 Stem Walls In Coastal High Hazard Areas, stem walls shall not be permitted.

In Coastal A Zones, stem walls enclosing areas below the design flood elevation shall not be permitted. Stem walls supporting a floor system above and backfilled with soil or gravel to the underside of the floor system above shall be permitted where a shallow foundation in accordance with Section 4.5.1.2 is permitted to support the wall. Stem walls shall be designed to transfer all vertical and lateral forces to the slab above and to the foundation elements below. The design shall consider all forces resulting from flooding, including wave action, debris impact, erosion, and local scour. Where soils are susceptible to erosion and local scour, stem walls shall have deep footings. The stem wall design shall consider all forces resulting from soil pressure behind the walls, including the effect of hydrostatic loads, and all live and dead surcharge loads from the slab above. Flood openings shall not be required in stem walls constructed in accordance with this section.

4.6 ENCLOSED AREAS BELOW DESIGN FLOOD ELEVATION

Enclosed areas below the DFE shall be permitted only where all of the following conditions are met:

1. Enclosure walls shall be designed and constructed in accordance with Section 4.6.1 and Section 4.6.2;
2. Enclosed areas shall be used solely for parking of vehicles, building access, or storage; and
3. Where stairways are located inside enclosed areas with breakaway walls, exterior doors shall be required at the entry at the top of the stairs.

4.6.1 Breakaway Walls Breakaway walls and other similar nonload bearing elements, including open-wood lattice work and insect screening, shall be designed and constructed to fail under base flood or lesser conditions, without imparting additional flood loads to the foundation or superstructure and without producing debris damage to the structure or adjacent structures. Breakaway walls and their connections shall be designed in accordance with the requirements of Section 5.3.3 of ASCE 7 *Minimum Design Loads for Buildings and Other Structures* (ASCE/SEI 2010). Attendant utilities and equipment shall not be mounted on, pass through, or be located along breakaway walls.

4.6.2 Openings in Breakaway Walls Openings to allow for the automatic entry and exit of floodwaters during design flood conditions shall be installed in breakaway walls. The minimum total net area of the openings shall meet the requirements of Section 2.7.2.1 or Section 2.7.2.2 and shall be installed in accordance with Section 2.7.3.

4.7 EROSION CONTROL STRUCTURES

Bulkheads, seawalls, revetments, and other erosion control structures shall not be connected to the foundation or superstructure of a structure. Bulkheads, seawalls, revetments, and other erosion control structures shall be designed and constructed so as not to direct floodwaters or increase flood forces or erosion impacts on the foundation or superstructure of any structure.

CHAPTER 5
MATERIALS

5.1 GENERAL

New construction and substantial improvements in flood hazard areas shall be constructed with flood damage-resistant materials below the elevations specified in Table 5-1. Flood damage-resistant materials shall have sufficient strength, rigidity, and durability to adequately resist all flood-related and other loads unless designed to break away or as permitted elsewhere in this standard.

Exposed structural and nonstructural construction materials, including connections, shall be capable of resisting damage, deterioration, corrosion, or decay because of precipitation, wind-driven water, salt spray, or other corrosive agents known to be present.

Structural and nonstructural construction materials, including connectors and fasteners, below the elevations specified in Table 5-1 shall be capable of resisting damage, deterioration, corrosion, or decay because of direct and prolonged contact with floodwaters associated with design flood conditions.

5.2 SPECIFIC MATERIALS REQUIREMENTS FOR FLOOD HAZARD AREAS

5.2.1 Metal Connectors and Fasteners Metal plates, connectors, screws, bolts, nails, and other fasteners exposed to direct contact by floodwater, precipitation, or wind-driven water shall be stainless steel or equivalent corrosion resistant material, or hot-dip galvanized in accordance with ASTM A123/A123M *Standard Specification for Zinc (Hot-Dip Galvanized) Coatings on Iron and Steel Products* (ASTM 2012d), ASTM A153/A153M *Standard Specification for Zinc Coating (Hot-Dip) on Iron and Steel Hardware* (ASTM 2009), ASTM A653/A653M *Standard Specification for Steel Sheet, Zinc-Coated (Galvanized) or Zinc-Iron Alloy-Coated (Galvannealed) by the Hot-Dip Process* (ASTM 2011a), or ASTM A924/A924M *Standard Specification for General Requirements for Steel Sheet, Metallic-Coated by the Hot-Dip Process* (ASTM 2010a).

Table 5-1 Minimum Elevation below Which Flood Damage-Resistant Materials Shall Be Used

Flood Design Class[a]	Minimum Elevation in Flood Hazard Areas, Relative to Base Flood Elevation (BFE) or Design Flood Elevation (DFE)	Minimum Elevation in Coastal High Hazard Areas and Coastal A Zones, Relative to Base Flood Elevation (BFE) or Design Flood Elevation (DFE)
1	DFE	DFE
2	BFE + 1 ft or DFE, whichever is higher	BFE + 1 ft or DFE, whichever is higher
3	BFE + 1 ft or DFE, whichever is higher	BFE + 2 ft or DFE, whichever is higher
4	BFE + 2 ft or DFE, or 500-year flood elevation, whichever is higher	BFE + 2 ft or DFE, or 500-year flood elevation, whichever is higher

[a] See Table 1-1 for Flood Design Class descriptions.

5.2.2 Structural Steel Steel pipe piles shall conform to ASTM A252/252M *Standard Specification for Welded and Seamless Steel Pipe Piles* (ASTM 2010b).

Steel H piles and steel sheet piling shall conform to ASTM A572/A572M *Standard Specification for High-Strength Low-Alloy Columbium-Vanadium Structural Steel* (ASTM 2012a) or ASTM A690/A690M *Standard Specification for High-Strength Low-Alloy Nickel, Copper, Phosphorus Steel H-Piles and Sheet Piling with Atmospheric Corrosion Resistance for Use in Marine Environments* (ASTM 2012b).

Rolled steel shapes other than H piles shall conform to ASTM A36/A36M *Standard Specification for Carbon Structural Steel* (ASTM 2008a), ASTM A572/A572M *Standard Specification for High-Strength Low-Alloy Columbium-Vanadium Structural Steel* (ASTM 2012a), and ASTM A992/A992M *Standard Specification for Structural Steel Shapes* (ASTM 2011b).

Cast steel shoes, where provided, shall conform to ASTM A148/A148M *Standard Specification for Steel Castings, High Strength, for Structural Purposes* (ASTM 2008b).

5.2.2.1 Corrosive Environments Structural steel exposed to direct contact with saltwater, salt spray, or other corrosive agents known to be present shall be hot-dipped galvanized after fabrication. Secondary components such as angles, bars, straps, and anchoring devices shall be stainless steel or hot-dipped galvanized after fabrication in accordance with Section 5.2.1.

5.2.2.2 Noncorrosive Environments In areas where salt spray and other corrosive agents are known not to be present, exposed structural steel either shall meet the requirements of Section 5.2.2.1 or shall be primed, coated, plated, or otherwise protected against corrosion due to direct contact with floodwaters, precipitation, or wind-driven water.

Secondary components such as angles, bars, straps, and anchoring devices shall be stainless steel or hot-dipped galvanized after fabrication in accordance with Section 5.2.1.

Damage to protective finishes and coatings caused by handling or installation shall be repaired using procedures that result in protection equivalent to the stated requirement.

5.2.3 Concrete Ingredients of concrete, including admixtures and reinforcing steel, quality of concrete, and the design and construction thereof shall comply with ACI 318 *Building Code Requirements for Structural Concrete and Commentary* (ACI

2014) with special consideration for requirements concerning durability, including protection from chlorides and sulfates found in a saltwater environment.

5.2.4 Masonry Materials used in masonry construction, including masonry units, mortar, grout, reinforcing steel and accessories; quality of masonry; and the design and construction thereof shall comply with ACI 530/ASCE 5/TMS 402, *Building Code Requirements and Specifications for Masonry Structures and Related Commentaries* (ACI 2013).

5.2.5 Wood

5.2.5.1 Preservative Treatment Where preservative treated wood is required elsewhere in this standard, preservative treatment shall be in accordance with AWPA U1, *Use Category System: User Specification for Treated Wood* (AWPA 2014) for the species, product, and end use or, alternatively, wood members shall be heartwood of naturally decay resistant species.

5.2.5.2 Members and Connections The design and construction of wood members and connections shall comply with *National Design Specification for Wood Construction* (AWC 2012).

5.2.6 Finishes Interior and exterior finishes and trim shall be flood damage-resistant material.

CHAPTER 6
DRY FLOODPROOFING AND WET FLOODPROOFING

6.1 SCOPE

This section addresses design and construction requirements for floodproofing new construction and substantial improvements in flood hazard areas.

Design of floodproofing measures shall take into consideration and account for the flood loads and combination of loads in Section 1.6 and for the nature of flood-related hazards; frequency, depth, and duration of flooding; rate of floodwater rise and fall; floodwater temperature; soil characteristics; flood-borne contaminants and debris; flood warning time; access to and from floodproofed areas; structure occupancy and use; and functional dependence.

6.2 DRY FLOODPROOFING

Dry floodproofing shall be accomplished through the use of flood damage-resistant materials and techniques that render the dry-floodproofed portions of a structure substantially impermeable to the passage of floodwater below the elevations specified in Table 6-1. Sump pumps shall be provided to remove water accumulated due to any passage of vapor and seepage of water during the flooding event. Sump pumps shall not be relied on as a means of dry floodproofing. All materials that are below the elevations specified in Table 6-1 shall conform to the requirements of Chapter 5, except materials on the interior of dry floodproofed portions of building.

6.2.1 Dry Floodproofing Limitations Dry floodproofing of nonresidential structures and nonresidential areas of mixed-use structures shall not be allowed unless such structures are located outside of High Risk Flood Hazard Areas, Coastal High Hazard Areas, and Coastal A Zones. Dry floodproofing of residential structures or residential areas of mixed-use structures shall not be permitted.

Dry floodproofing shall be limited to (1) where flood velocities adjacent to the structure are less than or equal to 5 ft/s during the design flood, and (2) if human intervention is proposed, where conformance with the limitations of Section 6.2.3 is provided.

6.2.2 Dry Floodproofing Requirements Dry floodproofed areas of structures shall

1. Be designed and constructed so that any area below the applicable elevation specified in Table 6-1, together with attendant utilities, equipment, and sanitary facilities, is flood resistant with walls that are substantially impermeable to the passage of water. Walls, floors, and flood shields shall be designed and constructed to resist hydrostatic, hydrodynamic, and other flood-related loads, including the effects of buoyancy resulting from flooding to the elevation listed in Table 6-1;
2. Have any soil or fill adjacent to the structure compacted and protected against erosion and local scour in accordance with Section 2.4; and
3. Have at least one door satisfying building code requirements for an exit door, or a door, window, or other opening meeting the criteria of the building code for an emergency escape and rescue opening, above the applicable elevation specified in Table 6-1, and capable of providing human ingress and egress during the design flood.

6.2.3 Limits on Human Intervention Dry floodproofing measures that require human intervention to activate or implement prior to or during a flood shall be permitted only when all of the following conditions are satisfied:

1. The flood warning time (alerting potential flood victims of a pending flood situation) shall be a minimum of 12 h unless the community operates a flood warning system and implements an emergency plan to ensure safe evacuation of flood hazard areas, in which case human intervention is allowed only if the community can provide a minimum flood warning time equal to or longer than the cumulative time
   (a) to notify persons responsible for installation of floodproofing measures,
   (b) for responsible persons to travel to structures to be floodproofed,
   (c) to install, activate, or implement floodproofing measures, and
   (d) to evacuate all occupants from the flood hazard area.
2. All removable shields or covers for openings such as windows, doors, and other openings in walls shall be designed to resist flood loads specified in Section 1.6.
3. Where removable shields are to be used, a flood emergency plan shall be approved by the authority having jurisdiction and shall specify, at a minimum, the following information: storage locations of the shields, the method of installation, conditions activating installation, maintenance of shields and attachment devices, periodic practice of installing shields, testing sump pumps and other drainage measures, and inspecting necessary material and equipment to activate or implement floodproofing. The flood emergency plan shall be posted permanently in at least two conspicuous locations within the structure.

Table 6-1 Minimum Elevation of Floodproofing—Flood Hazard Areas Other Than Coastal High Hazard Areas,[a] Coastal A Zones,[a] and High Risk Flood Hazard Areas[a]

Flood Design Class[b]	Minimum Elevation of Floodproofing[c] Relative to Base Flood Elevation (BFE) or Design Flood Elevation (DFE)
1	BFE + 1 ft or DFE, whichever is higher
2[d]	BFE + 1 ft or DFE, whichever is higher
3[d]	BFE + 1 ft or DFE, whichever is higher
4[d]	BFE + 2 ft or DFE, or 500-year flood elevation, whichever is higher

[a] Dry floodproofing is not allowed in Coastal High Hazard Areas, Coastal A Zones, and other High Risk Flood Hazard Areas.
[b] See Table 1-1 for Flood Design Class descriptions.
[c] Wet or dry floodproofing shall extend to the same level.
[d] Dry floodproofing of residential buildings and residential portions of mixed-use buildings shall not be permitted.

6.3 WET FLOODPROOFING

Wet floodproofing shall be accomplished through the use of flood damage-resistant materials and techniques that minimize damage to a structure during periods where the lower portion of the structure is inundated by floodwater. All materials in contact with floodwaters shall conform with the requirements of Chapter 5.

6.3.1 Wet Floodproofing Limitations on Use Wet floodproofing of enclosed areas below the elevations listed in Table 6-1 shall be limited to

1. Flood Design Class 1 structures,
2. Enclosures used solely for parking of vehicles, building access, or storage,
3. Structures that are functionally dependent on close proximity to water, and
4. Agricultural structures not included in Flood Design Class 1 that cannot be located elsewhere and that are used solely for agricultural purposes.

6.3.2 Wet Floodproofing Requirements Wet floodproofing for flood events up to and including the design flood shall be accomplished by

1. Use of techniques that minimize damage to the structure associated with flood loads,
2. Meeting the requirements of Section 2.7 or Section 4.6, depending on the flood hazard area, and
3. Installation of utilities, including plumbing fixtures, in conformance with the requirements of Chapter 7.

CHAPTER 7
ATTENDANT UTILITIES AND EQUIPMENT

7.1 GENERAL

Attendant utilities and equipment shall be located above the minimum elevation specified in Table 7-1 unless the attendant utilities and equipment are (1) specifically allowed below the minimum elevation in this chapter, and (2) designed, constructed, and installed to prevent floodwaters, including any backflow through the system, from entering or accumulating within the components.

Attendant utilities and equipment shall be installed and anchored to resist flood forces.

Attendant utilities and equipment shall be permitted in areas of structures that are dry floodproofed in accordance with Section 6.2.

Attendant utilities and equipment shall not be mounted on, pass through, or be located along breakaway walls.

Elevated exterior platforms for attendant utilities and equipment shall be supported on piles or columns, or cantilevered from or knee braced to the structure. If piles or columns are utilized, they shall be adequately embedded to account for erosion and local scour around the supports.

7.2 ELECTRICAL SERVICE

7.2.1 Service Conduits and Cables Electrical service conduits and cables below the Design Flood Elevation (DFE) shall be waterproofed or conform to the provisions of NFPA 70 *National Electrical Code* (NFPA 2011) for wet locations. Underground service conduits and cables shall be buried to a depth sufficient to prevent movement, separation, or loss due to erosion and local scour under design flood conditions.

7.2.2 Exposed Conduits and Cables Electrical conduits and cables emerging from underground shall be designed, constructed, and installed to withstand flood-related loads, including the effects of buoyancy, hydrodynamic forces, and debris impacts. Waterproofing or protective enclosures shall be provided for nonwaterproof conduits and cables extending vertically to elevated structures. The enclosures shall be fastened securely to the structure; however, protective enclosures and electrical conduits and cables shall not be fastened to walls, enclosures, or structures intended to break away under flood conditions.

Electrical conduits and cables and protective enclosures installed below the elevations specified in Table 7-1 shall be sealed to prevent the entrance of floodwaters into electrical conduits and electrical service components.

7.2.3 Electric Meters Electric meters shall be located above the elevation specified in Table 7-1 unless the connection between the meter and electric lines extending vertically from the meter is within a waterproof enclosure.

7.2.4 Panelboards, Disconnect Switches, and Circuit Breakers The panelboards, load centers, main disconnect switches, and all circuit breakers shall be located above and be accessible from above the elevation specified in Table 7-1. Panelboards, load centers, main disconnect switches, and circuit breakers that are located more than 5 ft above the floor, or above the height specified in code or by the applicable authority, shall have a platform installed to provide access.

7.2.5 Electric Elements Installed below Minimum Elevations Where electrical conduits and cables are located below the elevation specified in Table 7-1, they shall be installed so as to drain water away from panelboards, controllers, switches, or other electrical equipment in accordance with NFPA 70 *National Electrical Code* (NFPA 2011).

Where required to meet life safety provisions of the code, a minimum number of lighting circuits, switches, receptacles, and lighting fixtures operating at a maximum voltage of 120 volts to ground shall be permitted below the elevation specified in Table 7-1. Electrical wiring shall be suitable for submergence in water, and only submersible-type splices shall be used. Switches, receptacles, and fixtures shall conform to the provisions of NFPA 70 *National Electrical Code* (NFPA 2011) for wet locations and shall contain no fibrous components.

All circuits, switches, receptacles, fixtures, and other electrical components and equipment that are the minimum necessary to meet life safety requirements shall be permitted to be installed below the elevation specified in Table 7-1, provided they are energized from a common distribution panel located above and

Table 7-1 Minimum Elevation of Attendant Utilities and Equipment

	Locate Attendant Utilities and Equipment Above[b]	
Flood Design Class[a]	Minimum Elevation in Flood Hazard Areas Relative to Base Flood Elevation (BFE) or Design Flood Elevation (DFE)	Minimum Elevation in Coastal High Hazard Areas and Coastal A Zones, Relative to Base Flood Elevation (BFE) or Design Flood Elevation (DFE)
1	DFE	DFE
2	BFE + 1 ft or DFE, whichever is higher	BFE + 1 ft or DFE, whichever is higher
3	BFE + 1 ft or DFE, whichever is higher	BFE + 2 ft or DFE, whichever is higher
4	BFE + 2 ft, DFE, or 500-year flood elevation, whichever is higher	BFE + 2 ft, DFE, or 500-year flood elevation, whichever is higher

[a] See Table 1-1 for Flood Design Class descriptions.
[b] Locate attendant utilities and equipment above elevations shown unless otherwise provided in Chapter 7.

accessible from above the elevation specified in Table 7-1 and shall be supplied by branch circuits originating from ground-fault circuit-interrupter breakers.

Main supply lines, meters, and other exterior electrical components installed below the elevation specified in Table 7-1 shall be installed on a nonbreakaway vertical structural element on the landward, downslope, or downstream side of the structure.

7.3 PLUMBING SYSTEMS

For the purposes of this standard, plumbing systems shall include sanitary collection systems, rain runoff collection systems, sanitary facilities and plumbing fixtures, water supply systems (including hot water heaters and water conditioning equipment), and sewage disposal systems.

7.3.1 Buried Plumbing Systems Where installed underground, piping and plumbing systems providing service to a structure shall be buried to a depth sufficient to prevent movement, separation, or loss due to flooding and erosion under design flood conditions.

7.3.2 Exposed Plumbing Systems Plumbing systems and components emerging from underground shall be designed, constructed, anchored, and protected to withstand flood-related loads, including the effects of buoyancy, hydrodynamic forces, and debris impacts.

7.3.3 Plumbing Systems Installed Below Minimum Elevations Plumbing systems and components, including plumbing fixtures, shall be elevated above the elevation specified in Table 7-1. Where plumbing systems and components have openings below the elevation specified in Table 7-1, the openings shall be protected with automatic backwater valves or other automatic backflow devices. Devices shall be installed in each line that extends below the DFE to prevent release of sewage into floodwaters and to prevent infiltration by floodwaters into the plumbing. Redundant devices requiring human intervention shall be permitted.

7.3.4 Sanitary Systems Sanitary systems shall be designed to minimize infiltration of floodwaters into the systems and discharges from the systems into floodwaters. Vents and openings shall be above the elevation specified in Table 7-1. Sanitary system storage tanks shall be designed, constructed, installed, and anchored to resist at least 1.5 times the potential buoyant and other flood forces acting on an empty tank during design flood conditions. Tanks and piping shall be installed to resist local scour and erosion.

Sanitary systems that must remain operational during or immediately after the design flood or lesser floods shall be equipped with a sealed storage tank that is sized to store at least 150% of the anticipated sewage flow associated with occupancy during flood conditions and during subsequent periods of saturated soil when sewage will not percolate.

7.4 MECHANICAL, HEATING, VENTILATION, AND AIR CONDITIONING SYSTEMS

Fuel supply lines extending below the elevation specified in Table 7-1 shall be equipped with a float operated automatic control valve to shut off fuel supply when floodwaters rise above the elevation specified in Table 7-1.

Ductwork and ductwork insulation shall be at or above the elevation specified in Table 7-1 unless designed, constructed, and installed to resist all flood-related loads and to prevent floodwater from entering or accumulating within the ductwork.

Air intake openings and exhaust outlets shall be at or above the elevation specified in Table 7-1.

Tanks associated with mechanical, heating, ventilation, and air conditioning systems shall meet the requirements of Section 9.7.

7.5 ELEVATORS

Unless otherwise permitted in this section, all elevator components shall be located above the elevation specified in Table 7-1. Elevator components located below the elevation specified in Table 7-1 shall be constructed of flood damage-resistant materials and shall be capable of resisting physical damage due to design flood conditions.

Hydraulic elevators shall be permitted below the elevation specified in Table 7-1, but the electrical control panel, hydraulic pump, and tank shall be elevated above that elevation. Drainage shall be provided for the elevator pit. The hydraulic lines, hydraulic cylinders, and buffer springs shall be located to protect the lines from physical damage due to design flood conditions or painted or coated with galvanic or rust-preventive paint.

For traction elevator systems, the machine room containing the electric hoist motors and electrical control panel, or the equipment associated with traction elevators without machine rooms, shall be located above the elevation specified in Table 7-1. Elevator components located in the hoistway below the elevation specified in Table 7-1 shall be protected from physical damage due to design flood conditions.

Where there is the potential for an elevator cab to descend below the elevation specified in Table 7-1 during a flood event, the elevator shall be equipped with controls that will prevent the cab from descending into floodwaters.

7.5.1 Elevator Shafts Elevator shafts shall be designed to resist flood loads in accordance with ASCE 7 (ASCE/SEI 2010).

Elevator shafts are not required to have flood openings.

Elevator shafts in Coastal High Hazard Areas and Coastal A Zones are not required to have breakaway walls.

CHAPTER 8
BUILDING ACCESS

8.1 GENERAL

Stairways and ramps that are located below the elevations specified in Tables 2-1 and 4-1 shall be designed and constructed to

1. Resist flood-related loads specified in Section 1.6 and minimize transfer of flood-related loads to the structure and structure foundation; or
2. Break away during design flood conditions without causing damage to the structure, including the foundation; or
3. Be retractable, or be able to be raised to or above the elevations specified in Tables 2-1 and 4-1, provided the ability to be retracted or raised prior to onset conditions of flooding is not contrary to means of egress requirements of the code; and
4. Use materials that conform to Chapter 5 for those portions of stairways and ramps that are located below the elevations specified in Tables 2-1 and 4-1, including items such as gates and doors.

In flood hazard areas other than Coastal High Hazard Areas and Coastal A Zones, enclosures for stairways and ramps that extend below the elevations specified in Table 2-1 shall conform to the requirements for enclosures in Section 2.7. In Coastal High Hazard Areas and Coastal A Zones, enclosures for stairways and ramps that extend below the elevations specified in Table 4-1 shall conform to the requirements for enclosures in Section 4.6.

Elevators shall conform to the requirements of Section 7.5.

CHAPTER 9
MISCELLANEOUS CONSTRUCTION

9.1 GENERAL

Miscellaneous structures and construction governed by this chapter shall be designed to withstand all flood-related loads as defined in Section 1.6 of this standard. Structures shall be elevated as required in Section 2.3 or Section 4.4 or shall be designed and constructed in conformance with this chapter. Materials below the design flood elevation shall conform to Chapter 5. Utilities shall be governed by Chapter 7.

9.2 DECKS AND PORCHES

Decks and porches may be attached to or detached from an adjacent structure and are subject to the requirements in Section 9.2.1 or Section 9.2.2, as applicable. Attached and detached decks and porches that are permitted below the base flood elevation shall not be permitted to be enclosed with solid, rigid walls. Attached and detached decks and porches that are permitted below the base flood elevation are permitted to be enclosed with open-wood lattice work and insect screening.

Attendant utilities and equipment associated with decks and porches shall conform to Chapter 7.

9.2.1 Attached Decks and Porches Decks and porches that are structurally connected to a structure shall be designed to function as a continuation of the structure. Foundations for decks and porches shall conform to the foundation requirements of Section 1.5.3. The foundation system supporting the structure shall be designed to accommodate any increased loads resulting from the structurally connected deck or porch.

In flood hazard areas other than Coastal High Hazard Areas, Coastal A Zones, and other High Risk Flood Hazard Areas, structurally connected decks and porches are permitted below the elevations specified in Table 2-1.

In Coastal High Hazard Areas, Coastal A Zones, and other High Risk Flood Hazard Areas, the lowest horizontal structural members of the deck or porch shall be at or above the elevations specified in Table 4-1 and shall be supported on foundations in conformance with Chapter 4 or cantilevered from or knee braced to the structure or the foundation system supporting the structure.

9.2.2 Detached Decks and Porches In flood hazard areas other than Coastal High Hazard Areas, Coastal A Zones, and other High Risk Flood Hazard Areas, decks and porches that are not structurally connected to a structure are permitted below the elevations specified in Table 2-1. Decks, porches, and patios shall be anchored to remain in place during the base flood or their foundations shall conform to the requirements of Section 1.5.3.

In Coastal High Hazard Areas, Coastal A Zones, and other High Risk Flood Hazard Areas, decks and porches that are not structurally connected to a structure are permitted below the elevations specified in Table 4-1 if designed and constructed (1) to remain intact and anchored to remain in place during the base flood or their foundations shall conform to the requirements of Section 1.5.3, or (2) to be frangible and break away so as to minimize the debris capable of causing significant damage to any structure.

9.3 CONCRETE SLABS

In flood hazard areas other than Coastal High Hazard Areas and Coastal A Zones, and other High Risk Flood Hazard Areas, concrete slabs used as parking pads, enclosure floors, landings, decks, walkways, patios, and similar uses shall be permitted below the base flood elevation, provided the concrete slabs are structurally independent of the primary foundation systems of buildings or, if structurally connected, the main structure shall be capable of resisting any added flood loads and effects of scour due to the presence of the slabs.

In Coastal High Hazard Areas, Coastal A Zones, and other High Risk Flood Hazard Areas, concrete slabs used as parking pads, enclosure floors, landings, decks, walkways, patios, and similar uses shall be permitted beneath or adjacent to structures, provided the concrete slabs are designed and constructed

1. To be structurally independent of the primary foundation system of the structure, to not transfer flood loads to the main structure, and to be frangible and break away so as to minimize the debris capable of causing significant damage to any structure. Reinforcing of concrete slabs, including welded wire reinforcement, shall not be used so as to minimize the potential for concrete slabs being a source of debris. Slabs shall not have turned down edges, and maximum slab thickness shall not exceed 4 in.; or
2. To be self-supporting structural slabs capable of remaining intact and functional under base flood conditions, including expected erosion or uplift pressures, and the main structure shall be capable of resisting any added flood loads and effects of local scour due to the presence of the slabs.

9.4 GARAGES, CARPORTS, AND ACCESSORY STORAGE STRUCTURES

Garages, carports, and accessory storage structures that are attached to or detached from an adjacent structure are subject to the requirements of Section 9.4.1 or Section 9.4.2, as applicable. The floors of garages, carports, and accessory storage structures shall be at or above grade on at least one side. The foundations shall conform to the foundation requirements of Section 1.5.3.

In flood hazard areas other than Coastal High Hazard Areas, Coastal A Zones, and other High Risk Flood Hazard Areas, garages and accessory storage structures are permitted to be dry floodproofed or wet floodproofed in conformance with Chapter 6.

In Coastal High Hazard Areas, Coastal A Zones, and other High Risk Flood Hazard Areas, concrete slabs used as floors of garages, carports, and accessory storage structures shall conform to the requirements of Section 9.3.

Materials below the elevations specified in Table 5-1 shall conform to Chapter 5. Attendant utilities and equipment associated with garages, carports, and accessory storage structures shall conform to Chapter 7.

9.4.1 Attached Garages, Carports, and Accessory Storage Structures In flood hazard areas other than Coastal High Hazard Areas, Coastal A Zones, and other High Risk Flood Hazard Areas, attached garages, carports, and accessory storage structures that are structurally connected to a structure shall be designed to function as a continuation of the structure. Attached garages and accessory storage structures are permitted below the elevations specified in Table 2-1, provided the walls meet the requirements of Section 2.7.

In Coastal High Hazard Areas, Coastal A Zones, and other High Risk Flood Hazard Areas, attached garages, carports, and accessory storage structures that are elevated shall meet the requirements of Chapter 4. Attached garages, carports, and accessory storage structures are permitted below the elevation specified in Table 4-1, provided the walls comply with the requirements of Section 4.6.

9.4.2 Detached Garages, Carports, and Accessory Storage Structures In flood hazard areas other than Coastal High Hazard Areas, Coastal A Zones, and other High Risk Flood Hazard Areas, garages, carports, and accessory storage structures that are not structurally connected to a structure are permitted below the elevations specified in Table 2-1, provided the walls meet the requirements of Section 2.7.

In Coastal High Hazard Areas, Coastal A Zones, and other High Risk Flood Hazard Areas, detached garages, carports, and accessory storage structures that are elevated shall meet the requirements of Chapter 4. Detached garages, carports, and accessory storage structures are permitted below the elevation specified in Table 4-1, provided the walls comply with the requirements of Section 4.6.

9.4.3 Multistory Parking Structures Foundations of multistory parking structures shall conform to the foundation requirements of Section 1.5.3.

In flood hazard areas other than Coastal High Hazard Areas, Coastal A Zones, and other High Risk Flood Hazard Areas, multistory parking structures shall be permitted below the elevations specified in Table 2-1 provided (1) the tops of slabs of the lowest level are at or above grade on at least one side and any enclosed areas below the specified elevations conforms to the enclosure requirements of Section 2.7; or (2) any enclosed areas below the specified elevations are dry floodproofed in conformance with Chapter 6.

In Coastal High Hazard Areas, Coastal A Zones, and other High Risk Flood Hazard Areas, multistory parking structures shall be permitted below the elevations specified in Table 4-1 provided:

1. The parking structures are not structurally connected to buildings or other structures;
2. The tops of slabs of the lowest level are at or above grade on at least one side;
3. Columns, shear walls, beams, slabs, ramps, and vehicle barriers permit the free passage of floodwaters and waves through the structures;
4. Enclosed areas below the specified elevations, including stairwells and elevator shafts, conform to the enclosure requirements of Section 4.6 and Section 7.5, as applicable; and
5. Other components below the specified elevations are limited to lighting and utilities required for parking facilities, vehicle lifts, exit gates, controls, and pay stations.

9.5 CHIMNEYS AND FIREPLACES

In flood hazard areas other than Coastal High Hazard Areas, Coastal A Zones, and other High Risk Flood Hazard Areas, chimneys are permitted to extend below the elevation specified in Table 7-1. Chimneys extending below the elevation specified in Table 7-1 shall be supported vertically, shall be independent of the structure, and shall be designed and constructed to withstand all flood-related loads.

In Coastal High Hazard Areas, Coastal A Zones, and other High Risk Flood Hazard Areas, the base of the chimney or fireplace shall not extend below the elevation specified in Table 7-1. Where vertical support is required, chimneys shall be supported vertically on piles or column foundations, and foundation embedment shall be at least as deep as the rest of the structure foundation or deeper where needed to support the chimney against flood-related and other loads. The chimney or fireplace system shall be designed to minimize transfer of flood-related and other loads or load combinations to the structure and structure foundation.

9.6 POOLS

In-ground and aboveground pools shall be designed to withstand all flood-related loads and load combinations.

Mechanical equipment for pools such as pumps, heating systems, and filtering systems, and their associated electrical systems shall comply with Chapter 7.

Membrane structures over pools shall be permitted provided (1) the structural frame, where provided, supporting the membrane is designed to resist flood loads, and (2) the membrane is anchored to the frame, where provided, or the ground.

9.6.1 Pools in Flood Hazard Areas Other than Coastal High Hazard Areas and Coastal A Zones In flood hazard areas other than Coastal High Hazard Areas and Coastal A Zones, pools that are structurally connected to buildings or structures shall be designed to function as a continuation of the building or structure, which shall be designed to accommodate any increased loads resulting from the structurally connected pool.

9.6.2 Pools in Coastal High Hazard Areas, Coastal A Zones, and Other High Risk Flood Hazard Areas In Coastal High Hazard Areas, Coastal A Zones, and other High Risk Flood Hazard Areas:

1. Pools shall either
   a. Be elevated so that the lowest horizontal structural member is at or above the design flood elevation,
   b. Be designed and constructed to break away during design flood conditions without producing debris capable of causing significant damage to any structure, or
   c. Be designed and constructed to remain in the ground during design flood conditions without obstructing flow that results in damage to any structure.

2. Pools shall be located and designed to be structurally independent of buildings and structures unless the pools are located in or on elevated floors or roofs that are at or above the design flood elevation.

9.7 TANKS

Tanks and tank inlets, fill openings, outlets, and vents that are located below the design flood elevation shall be designed, constructed, installed, and anchored to resist all flood-related and other loads, including the effects of buoyancy, during flooding up to and including the design flood and without release of contents into floodwaters or infiltration of floodwaters into the tanks. Loads on underground tanks and aboveground tanks exposed to flooding shall be determined assuming at least 1.5 times the potential buoyant and other flood forces acting on the empty tank.

9.7.1 Aboveground Tanks In flood hazard areas other than Coastal High Hazard Areas, Coastal A Zones, and other High Risk Flood Hazard Areas, aboveground tanks shall be either

1. Elevated to or above the design flood elevation on platforms or structural fill,
2. Elevated to or above the elevation in Table 2-1 where attached to structures and the foundation system supporting the structures shall be designed to accommodate any increased loads resulting from the attached tanks,
3. Permitted below the design flood elevation where the tank and its foundation are designed to resist all flood-related loads including floating debris, or
4. Permitted below the design flood elevation where the tank and its foundation are designed to resist flood loads and are located inside a barrier designed to protect the tank from floating debris.

In Coastal High Hazard Areas, Coastal A Zones, and other High Risk Flood Hazard Areas, aboveground tanks shall be elevated to or above the design flood elevation on platforms that conform to the foundation requirements of Section 4.5. Aboveground tanks shall not be permitted to be located under elevated structures or attached to structures at elevations below the minimum elevations specified in Table 4-1.

9.7.2 Underground Tanks In Coastal High Hazard Areas, Coastal A Zones, and other High Risk Flood Hazard Areas, determination of flood-related loads shall take into consideration the eroded ground elevation.

CHAPTER 10
REFERENCES

All documents referenced in this standard are listed below with full titles and dates.

American Concrete Institute (ACI). (2013). "Building code requirements and specifications for masonry structures and related commentaries," *ACI 530/ ASCE 5/TMS 402*, Farmington Hills, MI.

ACI. (2014). "Building code requirements for structural concrete and commentary," *ACI 318*, Farmington Hills, MI.

American Society of Civil Engineers/Structural Engineering Institute (ASCE/SEI). (2010). "Minimum design loads for buildings and other structures," *ASCE/SEI 7*, Reston, VA.

American Society for Testing and Materials (ASTM). (2008a). "Standard specification for carbon structural steel." *A36/A36M*, ASTM International, West Conshohocken, PA.

ASTM. (2008b). "Standard specification for steel castings, high strength, for structural purposes." *A148/A148M*, ASTM International, West Conshohocken, PA

ASTM. (2009). "Standard specification for zinc coating (hot-dip) on iron and steel hardware." *A153/A153M*, ASTM International, West Conshohocken, PA.

ASTM. (2010a). "Standard specification for general requirements for steel sheet, metallic-coated by the hot dip process." *A924/A924M*, ASTM International, West Conshohocken, PA.

ASTM. (2010b). "Standard specification for welded and seamless steel pipe piles." *A252/A252M*, ASTM International, West Conshohocken, PA.

ASTM. (2011a). "Standard specification for steel sheet, zinc-coated (galvanized) or zinc-iron alloy-coated (galvannealed) by the hot-dip process." *A653/A653M*, ASTM International, West Conshohocken, PA.

ASTM. (2011b). "Standard specification for structural steel shapes." *A992/ A992M*, ASTM International, West Conshohocken, PA.

ASTM. (2012a). "Standard specification for high-strength low-alloy columbium-vanadium structural steel." *A572/A572M*, ASTM International, West Conshohocken, PA.

ASTM. (2012b). "Standard specification for high-strength low-alloy nickel, copper, phosphorus steel H-piles and sheet piling with atmospheric corrosion resistance for use in marine environments." *A690/A690M*, ASTM International, West Conshohocken, PA.

ASTM. (2012c). "Standard specification for round timber piles." *D25*, ASTM International, West Conshohocken, PA.

ASTM. (2012d). "Standard specification for zinc (hot-dip galvanized) coatings on iron and steel products." *A123/A123M*, ASTM International, West Conshohocken, PA.

ASTM. (2012e). "Standard test methods for laboratory compaction characteristics of soil using modified effort [(56,000 ft · lbf/ft^3 (2,700 kN · m/m^3)]." *D1557*, ASTM International, West Conshohocken, PA.

ASTM. (2012f). "Standard test methods for laboratory compaction characteristics of soil using standard effort [(12,400 ft · lbf/ft^3 (600 kN · m/m^3)]." *D698*, ASTM International, West Conshohocken, PA.

American Wood Council (AWC). (2012). "National design specification for wood construction." *AWC NDS*, Leesburg, VA.

American Wood Protection Association (AWPA). (2014). "Use category system: User specification for treated wood." *AWPA U1*, Selma, AL.

National Fire Protection Association (NFPA). (2011). "National electrical code." *NFPA 70*, Quincy, MA.

CHAPTER C1
GENERAL

(This commentary is not part of ASCE/SEI 24-14. It is included for information purposes.)

C1.1 SCOPE

The requirements of this standard are developed through a rigorous consensus process of the American Society of Civil Engineers (ASCE) and are intended to protect public safety and property. This standard is, in part, based on examination of damage after floods and is intended to meet or exceed the requirements of the National Flood Insurance Program (NFIP). Any conflicts or differences between this standard and other applicable regulations should be resolved such that compliance with NFIP requirements is equaled or exceeded. This standard is intended for use by registered design professionals, and designs based on this standard should be signed and sealed by the professional who is responsible for the design.

Certain terms not defined in the standard are defined in this commentary.

This standard applies to the design and construction of new buildings and structures located in flood hazard areas and subsequent work on buildings and structures that were designed and built to comply with requirements for flood hazard areas. In addition, the standard applies to foundations intended for modular buildings and to foundations for placement or replacement of manufactured housing units. Additional guidance on manufactured home foundations may be found in FEMA P-85, *Protecting Manufactured Homes from Floods and Other Hazards* FEMA (2009b).

This standard also applies to buildings and structures that are to be substantially improved or that incur substantial damage. The terms "substantial improvement" and "substantial damage" are defined in the standard. The authority having jurisdiction makes the determination as to whether work proposed for an existing building is substantial improvement or repair of substantial damage, see FEMA P-758, *Substantial Improvement/ Substantial Damage Desk Reference* (FEMA 2010c). Buildings and structures that are substantially improved or that are repaired after substantial damage are required to be brought into compliance with the requirements for new structures.

Work on existing buildings that is subject to this standard includes additions that change the footprint of a structure and additions that change the number of floors. Work also includes alteration or replacement of foundations, an activity that may be undertaken specifically to improve a structure's resistance to flood loads and damage [e.g., to raise a substantially improved or substantially damaged structure above the design flood elevation (DFE)].

This standard is not intended to apply to routine or minor repairs or improvements to a building or structure, nor does it apply to existing buildings that are not substantially improved or substantially damaged unless the authority having jurisdiction has adopted definitions of those terms that differ from the definitions in this standard. For example, some communities adopt a "cumulative" provision that tracks permits over time, and thus may have a different threshold for determining substantial improvement or substantial damage. Similarly, some communities adopt a "repetitive loss" provision that tracks damage (typically only flood damage) over time, and thus may have a different threshold for substantial damage.

Even if compliance with this standard is not required, designers and owners of buildings located in flood hazard areas are encouraged to use the standard to evaluate measures that may reduce vulnerability to future floods. For example, upon examination and evaluation of flood loads, any building (even residential occupancies) may be retrofit with dry floodproofed measures, even if the resulting building is not in full compliance. Consideration of such measures should account for the same factors that are covered in this standard. There are many reported cases where existing buildings have been successfully dry floodproofed to reduce damage from shallow flooding. However, designers are specifically cautioned that if the work (and all other work being done) is determined to be substantial improvement, then the authority having jurisdiction will require full compliance with this standard. As indicated in Chapter 6, residential buildings and residential portions of nonresidential buildings shall not be dry floodproofed if they are required to comply with this standard.

The provisions of this standard may not apply to certain "historic structures" as defined in this standard (see Fig. 1-1). The authorities having jurisdiction over historic structures, such as the State Historic Preservation Office and certain local historic preservation authorities that are approved by state programs or by the Secretary of the Interior, should be consulted to determine if proposed work on an historic structure allows the structure to retain its historic designation. If the structure will not retain its historic designation and if the work is determined to be substantial improvement or repair of substantial damage, then the provisions of this standard apply. If the standard does not apply, owners of historic structures are encouraged to use the standard to the extent practicable in order to reduce the exposure of historically significant buildings and their contents to future flood damage.

Fig. 1-2 illustrates how the standard is organized. Designers should determine which sections are specifically applicable on the basis of the type of flood hazard area. Chapter 1 contains provisions that apply to all new construction and substantial improvements in all flood hazard areas.

Chapter 2 applies to buildings and structures in flood hazard areas including High Risk Flood Hazard Areas other than Coastal High Hazard Areas and Coastal A Zones.

Chapter 3 applies in High Risk Flood Hazard Areas and basic provisions are outlined, but additional design requirements for these areas are not specified.

Chapter 4 applies to buildings and structures in Coastal High Hazard Areas and Coastal A Zones.

Chapter 5 (materials), Chapter 7 (attendant utilities and equipment), and Chapter 8 (building access) apply in all flood hazard areas.

Chapter 6 applies to buildings and structures that are dry floodproofed or wet floodproofed.

Chapter 9 applies to miscellaneous construction such as decks and porches, concrete slabs used for nonstructural purposes, garages and accessory storage structures, chimneys and fireplaces, pools, and tanks.

This standard is not intended to preclude construction of piers, docks, wharves, and other water-dependent (functionally dependent) structures.

C1.2 DEFINITIONS

500-Year Flood Elevation—This term refers to the water surface elevation of the 0.2% annual chance flood, commonly called the "500-year flood." Some Flood Insurance Rate Maps show and designate these areas as the "shaded" X Zone. This standard has specific requirements based on the 500-year flood elevation that apply to Flood Design Class 4 buildings and facilities that have critical or essential functions. Where 500-year flood elevations were determined during Flood Insurance Studies, the information is available in the studies that may be obtained at FEMA's Map Service Center (http://www.msc.fema.gov/). Where a flood hazard study has not developed 500-year flood elevations, further investigation should be undertaken to determine if a detailed hydraulic study should be performed.

A Zone—Common term used to refer collectively to areas within special flood hazard areas shown on Flood Insurance Rate Maps (FIRMs) that are not subject to high velocity wave action and which are designated with the letter "A," including Zones A, AE, A1-30, A99, AR, AO, and AH. The term "unnumbered A Zone" is sometimes used to refer to zones that do not have base flood elevations (BFEs). (See also Section C1.3).

Accessory Storage Structure—The NFIP allows communities to approve certain buildings without requiring elevation if those buildings are used only for storage or parking of vehicles (detached garages). With this definition, it is clearer that accessory structures intended for other purposes are subject to the elevation requirements of the standard. FEMA guidance for accessory structures suggests that communities establish a dollar-value threshold, which would influence the size of such structures.

Basement—This standard defines "basement" consistent with the NFIP and the definition in building codes that is specific for application in flood hazard areas. To avoid classification as a basement, designers should specify that the interior grade or interior floor is at or above the finished exterior grade along one entire side of a building.

Breakaway Wall—By using the phrase "under base flood or lesser flood conditions" in the definition, this standard is not requiring walls to fail under any applied load that is less than that occurring during the base flood. The load requirements of ASCE 7 (Standard Reference, ASCE/SEI 2010) must be met. ASCE 7 requires breakaway walls to withstand the design lateral wind loads and seismic loads but to fail under the specific lateral loads associated with base flood conditions. In no instance does ASCE 7 permit breakaway walls that will fail at lateral loads less than 10 lb/ft^2.

Coastal A Zone—Coastal A Zone designation is used to facilitate application of load combinations contained in ASCE 7 (Standard References, ASCE/SEI 2010). Coastal A Zones are areas where wave heights greater than or equal to 1.5 ft during the base flood are expected (Fig. C1-1); they lie landward of zones designated by FEMA as V Zones and landward of an open coastal shoreline where V Zones have not been mapped. Post-flood field investigations and laboratory testing indicate significant structural damage occurs in these areas and is associated with wave impacts. Some Coastal A Zones experience erosion that is similar to that experienced in V Zones, and this standard requires design and construction similar to that required in V Zones.

The Coastal A Zone is not delineated by FEMA as a separate zone on FIRMs. Instead, the Coastal A Zone is determined by the Limit of Moderate Wave Action (LiMWA) delineated by FEMA on revised FIRMs prepared after December 2009 (FEMA 2008e, 2013b). Some communities separately designate Coastal A Zones by other means, such as by a geographic feature or a distance from the V Zone boundary or shoreline.

For a Coastal A Zone to be present, two conditions are required: (1) a stillwater depth greater than or equal to 2 ft; and (2) wave heights greater than or equal to 1.5 ft. Note that the stillwater depth requirement is necessary but is not sufficient by itself to render an area a Coastal A Zone. Many flood hazard areas have stillwater flood depths in excess of 2 ft but may not experience wave heights greater than or equal to 1.5 ft.

Design Flood Elevation (DFE)—Based on the design flood, the DFE is the higher of the base flood elevation (BFE) shown on FIRMs prepared by FEMA or the flood elevations shown on the map adopted by a community. Most communities use FIRMs, and thus the DFE is equal to the BFE. Communities may elect to adopt flood elevations that are higher than those determined by FEMA. Such higher flood elevations may be adopted for a number of reasons, for example, to show future conditions (assuming predicted upland development, subsidence, or sea level rise), to reflect the flood of record or other flood events that exceeded the 1% annual chance flood, or to incorporate freeboard as an additional factor of safety to reflect local conditions (such as high debris loads not reflected in BFEs).

Development—This term refers to any artificial change to improved or unimproved real estate, including but not limited to structures, permanent storage of materials, mining, dredging, filling, grading, paving, excavations, drilling operations, and other land disturbing activities. Although only the design and construction of buildings and structures is subject to this standard, it is valuable for users to understand that development is broadly defined by the NFIP and regulated by communities that participate in the program. See commentary for Section 2.2 and the definition of floodplain management ordinance.

Essential Facility—See Table 1-1, Flood Design Class 4. Building codes also define "essential facility," which is used by the codes for occupancy category IV or risk category IV structures.

Federal Emergency Management Agency (FEMA)—Federal agency, or part of a federal agency, that, in addition to carrying out other activities, oversees the administration of the NFIP. FEMA is charged with coordinating with states and localities to prepare for, mitigate the effects of, respond to, and recover from major domestic disasters, both natural and human induced.

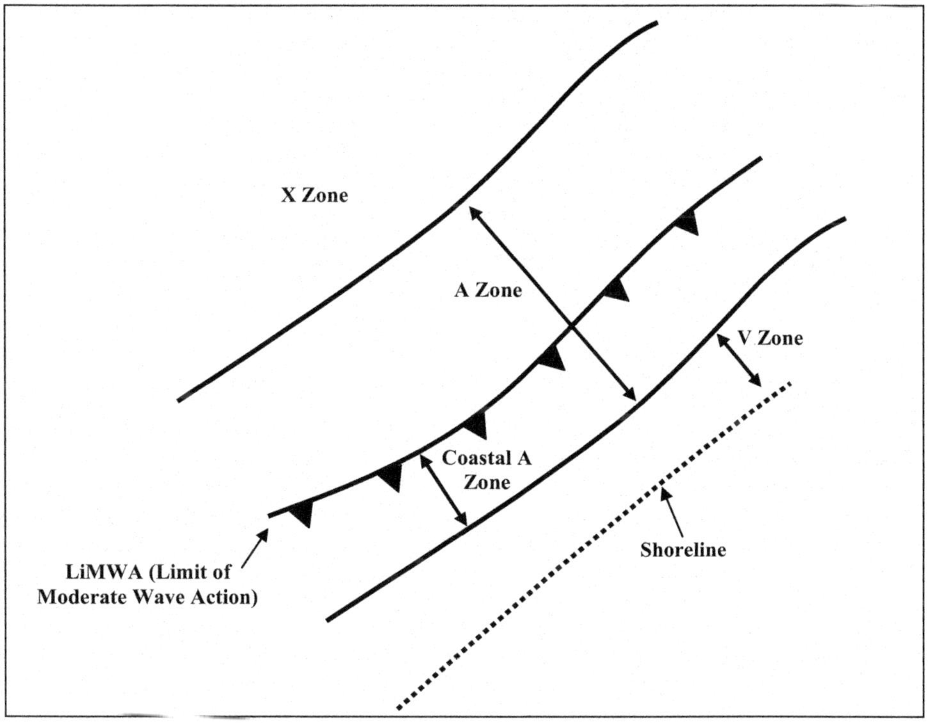

Fig. C1-1. Schematic of the Coastal A Zone

Flash Flood—Flood hazard areas subject to conditions that meet this definition are not separately identified on FIRMs or identified in Flood Insurance Studies. These conditions generally occur in areas where there is considerable topographic relief, such as steeply sloping waterways in mountainous or hilly areas. Flash flooding can also occur in waterways when intense rains fall on headwater areas, particularly in arid regions. Designers may need to consult with local authorities or water resources professionals to determine if historical events or anticipated flooding conditions meet this definition. See commentary for Section 3.3.

Flood Design Class—This standard requires all buildings and structures within its scope to be assigned a Flood Design Class (see Section C1.4.3). This is a significant change from previous editions of this standard, which used the occupancy category from earlier editions of ASCE 7.

Flood Hazard Boundary Map (FHBM)—First flood risk map prepared by FEMA for a community, which identifies flood hazard areas based on the approximation of land areas in the community having a 1% or greater chance of flooding in any given year.

Flood Insurance Rate Map (FIRM)—Official map on which FEMA has delineated both the special flood hazard areas and the risk premium zones applicable to a community. Many communities have multiple panels comprising the FIRM, and panels may have different effective dates. Users should verify that the most current effective map is used in determining flood hazard areas, floodway areas, and BFEs. FEMA's online Map Service Center is accessible at http://www.msc.fema.gov. Many states and communities also provide online access to FIRMs.

Flood Insurance Study (FIS)—A report prepared by FEMA to document the examination, evaluation, and determination of flood hazards and, if appropriate, corresponding water surface elevations, or an examination, evaluation, and determination of mudslide (mudflow) or flood-related erosion hazards. The FIS typically includes a section on discharges of streams studied, water surface elevation profiles for floods of different frequencies, and floodway data tables, which also include some velocity information. For coastal communities, the FIS typically includes water surface elevations for floods of different frequencies at identified locations (transects) along the shoreline.

Floodplain Management Ordinance—Ordinance adopted by the authority having jurisdiction to regulate development in flood hazard areas. The ordinance addresses all development, including development activities that are outside the scope of this standard.

Floodproofing—Measures outlined in this standard may apply to new construction or improvement or retrofit of existing structures. Floodproofing is intended to protect areas within a structure that are below specific elevations. The standard specifies limitations on the use of dry floodproofing and wet floodproofing.

Freeboard—Additional height used as a factor of safety in setting the minimum elevation of a structure, or floodproofing measures applied to a structure, to provide a higher level of protection for certain structures based on importance (Flood Design Class). Freeboard also is used to compensate for factors that may increase flood heights and for uncertainties inherent in determining flood frequencies and flood elevations. States and communities may adopt freeboard requirements that exceed the elevation requirements of this standard.

Functionally Dependent—Defined in the standard and used in Section 6.3.1 where certain structures in close proximity to water may be wet floodproofed. Designers should note that many other structures that are typically located near water (such as marina offices, chandlers, and bait and tackle shops) are not specified in the definition. The definition in the standard is the same as defined by the NFIP (44 CFR Part 59.1).

Lowest Adjacent Grade—Lowest elevation of the natural or regraded ground surface, or structural fill, at the location of a structure.

Mixed Use—Any building or structure that has portions that are classified nonresidential and portions that are classified residential. See commentary for the term "residential portions of mixed-use buildings." Dwellings and dwelling units are not considered mixed use; they are residential. A dwelling is a building that contains one or two dwelling units. A dwelling unit provides for living, sleeping, eating, cooking, and sanitation, thus providing a complete independent living arrangement for one or more persons.

Mudslide—Definition is consistent with the NFIP use. However, it is intended to include only classes of flow that are technically recognized as mudflow and mud flood. Mudslides (mudflows and mud floods) occur where sediment concentrations are between 0.20 and 0.55 (by volume) or between 0.41 and 0.79 (by weight). Where sediment concentrations are less than those limits, the condition is a flood. Where sediment exceeds those limits, the condition is a landslide. Landslides are not intended to be covered in this standard, as they are not directly flood related.

National Flood Insurance Program (NFIP)—Federal program that enables property owners in participating communities to purchase federally backed flood insurance as financial protection against flood losses in exchange for state and community floodplain management regulations that reduce future flood damage. The program is administered by FEMA.

New Construction—As defined by this standard and the NFIP, this term does not simply distinguish between new buildings and existing buildings (as defined by building codes). It is used to separate buildings built before communities adopted floodplain management requirements from those built after. In part, use of this term is intended to ensure that subsequent work performed on a building that meets the definition will be conducted in compliance with the requirements of the standard and thus not violate any aspect of the building that was required for compliance at the time the building was designed and built. It is common, but misleading, to use the terms "pre-FIRM" and "post-FIRM" to distinguish between new construction and buildings that pre-date when a community adopted its first floodplain management requirements. Especially as FEMA revises FIRMs, sometimes changing flood zones and BFEs, reliance on those terms can lead to incorrect interpretations.

Prolonged Contact with Floodwaters—Definition used in this standard is consistent with FEMA Technical Bulletin 2, *Flood Damage-Resistant Materials Requirements for Buildings Located in Special Flood Hazard Areas* (FEMA 2008b). Duration of flooding varies significantly based on many factors. Flooding in relatively smaller watersheds and coastal flooding usually are of short duration, while flooding in large watersheds may last for weeks or even months. Where durations are known or determined through analyses to be longer than 72 h, designers should be aware that some materials identified as "flood damage-resistant" may perform acceptably for longer than 72 h, but others may have limited flood resistance beyond 72 h. Designers should consider specifying materials that resist flood damage for longer durations in these cases.

Residential portions of mixed-use buildings—This standard defines "residential" and "nonresidential" and uses the term "mixed use," which is defined in this commentary. Residential portions of mixed-use buildings include residential units, dwelling units, and other spaces meeting the definition of "residential." Congregate living arrangements, such as nursing homes, assisted living, dormitories, group homes, sororities, or fraternities, where residents share living, bathroom, or kitchen space are also considered residential.

V Zone—Common term used to refer collectively to Coastal High Hazard Areas shown on FIRMs that are subject to high velocity wave action and that are designated with the letter "V," including Zones V, VE, V1-30, and VO. The term "unnumbered V Zone" is sometimes used to refer to velocity zones that do not have BFEs.

Wave Crest Elevation—Elevation of the crest of a wave, usually referenced to North American Vertical Datum of 1988 (NAVD), National Geodetic Vertical Datum of 1929 (NGVD), or another datum.

C1.3 IDENTIFICATION OF FLOOD HAZARD AREAS

The national basis for floodplain management is the 1% annual chance flood (commonly called 100-year flood). This event has a 1% chance of being equaled or exceeded in any given year. The NFIP refers to this flood as the "base flood," which is shown on FIRMs or Flood Hazard Boundary Maps (FHBMs). A base flood elevation (BFE)—the water surface elevation associated with the base flood relative to NAVD, NGVD, or another datum—is identified on many FIRMs.

The determination as to whether a building site or structure lies within flood hazard areas begins by determining the flood map that has been adopted by the community. Although most communities adopt flood hazard maps prepared by FEMA, other maps that show larger flood-affected areas, areas of historical flooding, or higher design flood elevations (DFEs) may be used for regulatory purposes (see commentary for "design flood").

In flood hazard areas for which detailed elevation information is not shown on a community's FIRMs, information may be obtained from technical studies produced in conjunction with FEMA mapping or from other sources, such as the U.S. Army Corps of Engineers, the U.S. Natural Resources Conservation Service (formerly known as the U.S. Soil Conservation Service), the U.S. Geological Survey, and state or local floodplain management agencies and transportation agencies.

In the absence of adequate information from the listed organizations, practitioners in hydrology, hydraulics, or coastal engineering can be consulted to determine the flood hazard potential for a site, with an analysis of site characteristics, with engineering calculations performed with a variety of FEMA-approved models (listed by FEMA at http://www.fema.gov/national-flood-insurance-program-flood-hazard-mapping/numerical-models-meeting-minimum-requirement-0), and with historical flood records (if available). Practitioners in hydrology and hydraulics can calculate riverine flood elevations and floodways. Practitioners in coastal engineering can calculate storm surge elevations and storm-induced erosion with a variety of approved models. It should be noted that in areas where infrequent but extreme flood events occur, the lack of sufficient historical data makes such determinations difficult and reduces the reliability of flood frequency analyses.

In flood hazard areas subject to flooding from riverine sources (noncoastal), flood zones begin with the letter "A." Many waterways that have detailed elevations also have designated "floodways." Floodways generally are areas where the greatest conveyance of floodwaters occurs, and they typically have greater depths and higher velocities than do areas outside the floodway, sometimes called the "floodway fringe" (Fig. C1-2). Development within floodways can increase flood elevations; thus, additional engineering evaluation of the impact of proposed development is required (see Section C1.5.4).

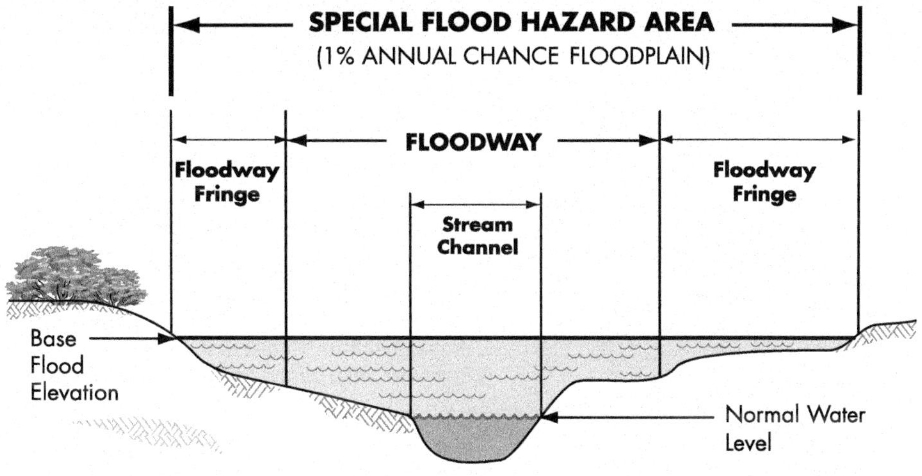

Fig. C1-2. Floodway schematic
Source: U.S. Government FEMA

In flood hazard areas subject to flooding from coastal sources, zones beginning with the letter "A" represent areas that have a 1% or greater probability of flooding during any year but that are not expected to experience wave heights of 3 ft or more or wave runup depths of 3 ft or greater (high velocity wave action). Zones beginning with the letter "V" represent areas that have a 1% or greater probability of flooding during any year and where high velocity wave action is anticipated during the 1% flood. Zones beginning with a letter "V" are also referred to as Coastal High Hazard Areas; however, high velocity wave action can occur in lake or riverine floodplains under certain circumstances as well. Fig. C1-3 illustrates flood zones identified on FIRMs.

Coastal A Zones, where wave forces and the potential for erosion are present, are described in commentary for the definition (also see Limit of Moderate Wave Action) in Section C2.1.

Designers are reminded to exercise caution regarding the use of the term "100-year flood." The public frequently misinterprets the term, believing that the occurrence of a 100-year flood precludes another flood of similar magnitude for 100 years. Wherever possible, the term "1% annual chance flood" should be used instead of "100-year flood."

Approximately 22,000 communities participate in the NFIP by adopting floodplain management regulations or a combination of regulations and building codes. Communities that participate in the NFIP must meet or exceed the minimum NFIP requirements. Designers should be aware, however, that many states and communities enforce more restrictive requirements. More restrictive requirements may be found in virtually all aspects of local regulations. Some of the more common are the following: adopting a higher flood elevation than the NFIP's BFE; requiring construction to be elevated higher than the NFIP's requirements; requiring engineered foundations in all flood hazard areas; limiting the size of enclosed areas under elevated buildings; or imposing other more restrictive siting, development, or construction standards. The authority having jurisdiction should be consulted to determine the applicable requirements for construction in flood hazard areas.

For the purposes of this standard, the terms "design flood" and "design flood elevation" (DFE) will be used to refer to the locally adopted regulatory flood and its associated water surface elevation. If the authority having jurisdiction regulates to NFIP minimum requirements, then the design flood and DFE will be identical to the base flood and BFE. If the authority having jurisdiction regulates to a higher flood elevation, then the design flood and DFE will be greater than the base flood and BFE.

It should be noted that—in coastal areas—the DFE and BFE include the effects of waves; in other words, the DFE and BFE will be established at the wave crest elevation or wave runup elevation, not the stillwater elevation. The designer is referred to Chapter 3 for more information on this topic.

Even if the authority having jurisdiction has adopted NFIP minimum requirements, it may still be a prudent practice to examine whether sites near mapped flood hazard areas are subject to flooding by more severe floods than the base flood and to design and construct some buildings and structures to a higher elevation and more conservative standard than the minimum requirements. This is particularly pertinent when essential facilities (Flood Design Class 4) and other buildings of importance to the community (Flood Design Class 3) are proposed to be located in or near flood hazard areas. Careful consideration should be given to locating such facilities at locations where access and functioning are less likely to be impaired. If location in a flood hazard area is unavoidable, the designer is encouraged to apply a higher standard of care than what is represented by only fulfilling the minimum requirements.

In some cases mapping efforts are hindered by the fact that it is not possible to assign an accurate exceedance frequency to historical flood events, particularly in cases where an area is subject to unique hazards (alluvial fan flooding, mudflows, ice jams, debris blockage of culverts and bridges, etc.). Designers are cautioned that conditions associated with High Risk Flood Hazard Areas may be present but not identified on FIRMs, and the absence of such designation is not evidence that the conditions are not present. In addition, flood hazard maps are based on conditions at the time the flood study was performed and may not reflect conditions that have changed since that time, such as upland development that changes runoff characteristics, subsidence, riverine channel changes, shoreline erosion and accretion, sea level change, changes in precipitation patterns, changes associated with bridges, culverts, fills, and other artificial and natural alterations of the floodplain.

Due diligence is required to evaluate site conditions and research the nature and history of flooding and whether any of the characteristics associated with high risk areas are present or if the maps do not reasonably reflect current conditions.

FEMA Flood Insurance Rate Map (Coastal)

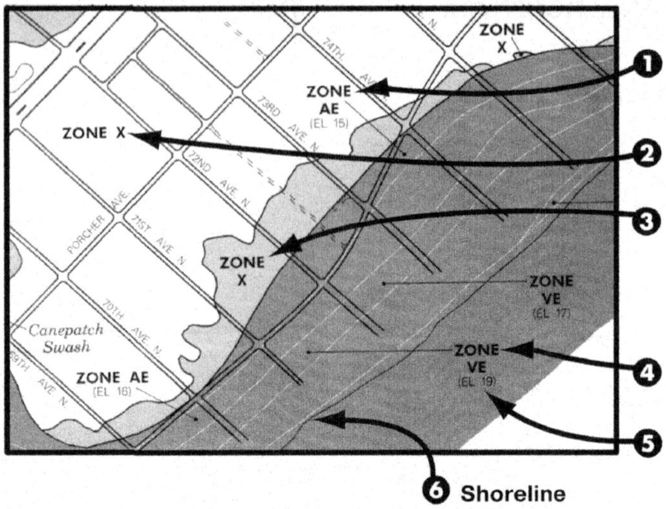

COASTAL FLOOD HAZARD ZONES

1. **Zone A** and **Zone AE** are subject to flooding by the base or 100-year flood (1% annual chance), and waves less than 3 feet (formerly called Zones A1-A30).
2. **Unshaded Zone X** is the area of minimal flood risk outside the 500-year floodplain, formerly called Zone C.
3. **Shaded Zone X** is subject to flooding by the 0.2-percent-annual-chance (500-year) flood and the 1-percent-annual-chance (100-year) flood with average depths of less than 1 foot or with drainage areas less than 1 square mile; also o designates areas protected from the 1-percent-annual-chance (100-year) flood by levees.
4. **Zone V** and **Zone VE** are where waves are expected to be 3 feet or more.
5. **Base Flood Elevation (BFE)** is the estimated water surface elevation (in feet above datum).

FEMA Flood Insurance Rate Map (Riverine)

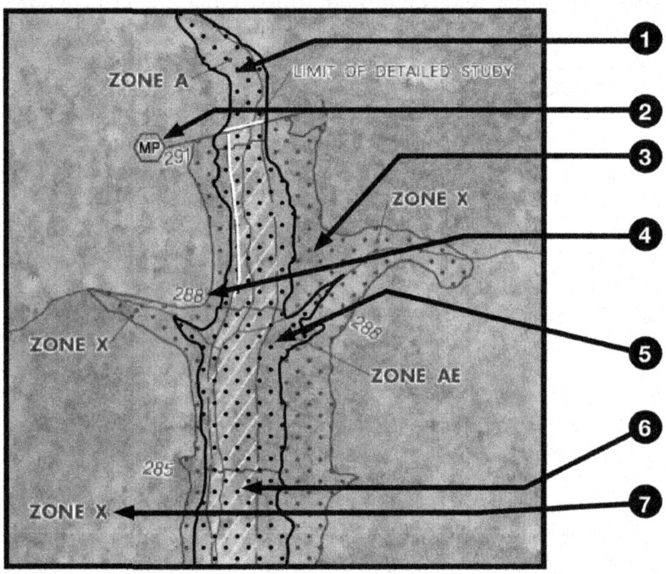

1. **Zone A** (approximate) is the flood hazard area without BFEs.
2. **Cross Section** location.
3. **Shaded Zone X** is the 0.2% annual chance (500-year) floodplain (formerly Zone B).
4. **Base Flood Elevation (BFE)** is the water surface elevation of the base flood rounded to the nearest whole foot (consult FIS profiles and tables for more accurate elevations).
5. **Zone AE** is the 1% annual chance (100-year) floodplain with BFEs (formerly Zones A1-A30).
6. The **Floodway** is the cross-hatched area.
7. **Unshaded Zone X** is all other areas considered low risk (formerly Zone C).

Flood Zones

Zone A identifies areas of special flood hazard subject to flooding associated with the 1%-annual-chance flood event in riverine areas and in coastal areas not subject to high velocity wave action.
Zone V identifies those portions of special flood hazard areas along open coasts that are subject to high velocity wave action from storms or seismic sources. Also called "coastal high hazard areas," Zone V areas are subject to additional hazards associated with storm wave heights that are predicted to be 3 ft or greater during 1%-annual-chance flood conditions.
Shaded Zone X identifies areas subject to flooding by the 0.2%-annual-chance (500-year) flood; areas subject to flooding by the 1%-annual-chance (100-year) flood with average depths of less than 1 ft or with drainage areas less than 1 sq mi; and areas protected from the 1%-annual-chance (100-year) flood by levees.
Unshaded Zone X identifies areas of minimal flood risk that are outside the 0.2%-annual-chance (500-year) floodplain.

Fig. C1-3. Flood zones on FIRMs

Source: U.S. Government FEMA

C1.4 IDENTIFICATION OF FLOOD-PRONE STRUCTURES

C1.4.1 General Flood hazard determinations should be based on the most recent and accurate maps, studies, and data available. Therefore, the community should always be contacted to obtain the latest information. However, the designer is warned that even the current effective maps and studies may be incomplete, inaccurate, or obsolete. In instances where maps, studies, or other information are known by the designer to be incomplete, inaccurate, or obsolete, the community may be able to assist the designer in obtaining and reviewing flood hazard data from other sources or may be able to provide a listing of other sources that the designer can contact.

This standard applies to all buildings and structures that are located, in whole or in part, in flood hazard areas. A building or structure that "straddles" a line between flood hazard areas, a floodway boundary line, or between a flood hazard area and an area that is subject to less frequent flooding, must meet the most restrictive requirements of the two areas. Designers are cautioned about imbuing a designation on a map with a higher degree of precision than intended. This caution is made with regard to several aspects of flood hazards and flood hazard maps, including but not limited to the following:

1. Determining flood hazard areas and flood elevations involves a great deal of analysis of historical records and/or modeling; the modeler has numerous assumptions to make and parameters to determine based on judgment.
2. The 1% annual chance flood is a statement of probability that is, for the most part, a function of past flood events; as the record of flood events at any given location accumulates and especially as large floods occur, a statistical analysis of that record likely will change the flood flows and elevations.
3. In riverine areas, changes in land use over time, development in upstream portions of watersheds, will alter the rainfall-runoff relationship and contribute to increases in flood elevations.
4. In coastal areas, modeled results are highly sensitive to the shoreline topography used, yet that topography may change over time because of erosion or development.
5. On FIRMs that are more than 10 to 15 years old, the base maps on which many flood hazard areas were delineated may not be of sufficient scale or accuracy to capture all ground variations; an area may not be shown as subject to flooding, but the elevation of the ground may, in fact, be lower than the design flood elevation.
6. The scale of most flood hazard maps is such that the width of lines can be a factor in determining whether a structure is "in" or "out" of the flood hazard area; designers are advised to be cautious in their determinations in order to protect life, property, and public safety.

C1.4.2 Consideration for Flood Protective Works Flood protective works, such as levees, dams, and floodwalls, do not provide complete or absolute protection to an area; they only reduce the frequency or intensity of flooding.

Thousands of miles of levees and other flood protective works are in existence. Many (but not all) have been designed, constructed, and maintained so as to provide protection against flood conditions, although perhaps not the design flood as defined by this standard. Many flood protective works were not designed and constructed to specific standards but were built for agricultural, emergency, or other purposes and may provide protection against only smaller floods (e.g., 5- or 15-year flood frequencies). Many will not provide protection during the base flood or the design flood and have not been recognized and credited by FEMA on the FIRMs or by other authorities on other flood hazard maps as providing flood protection. Unfortunately, levees and other similar works—even small and poorly maintained ones—often provide a false sense of security. When these works fail or are overtopped by floods that exceed their crest elevation, rapid inundation and high velocity flows often result. Even if the works do not fail or overtop during a flood, if not appropriately designed and maintained, they may cause flooding by trapping rainfall, runoff, and other waters behind them. See Chapter 4 of FEMA 116, *Reducing Losses in High Risk Flood Hazard Areas: A Guidebook for Local Officials* (FEMA 1987) for more information on areas behind unsafe or inadequate levees.

Only certain flood protective works that have been designed, constructed, and maintained to rigorous standards are recognized by FEMA as providing protection during the 1% annual chance flood event and are considered in floodplain mapping. When recognized by FEMA, the area protected by major flood control works is removed from the special flood hazard area, and mandatory purchase of federal flood insurance is not enforced by regulated mortgage lenders. Nevertheless, the area may be identified on the Flood Insurance Rate Map as having residual flood risk from floods greater than the 1% annual chance flood.

However, even certified flood protective works may not render the area behind them as being safe for the design and construction of structures at grade. For example, the integrity of certain flood protective works may have been reduced by flooding, erosion, poor maintenance, or other events since the floodplain was mapped. More commonly, areas protected by even well-designed and constructed works are known to suffer from drainage problems, especially during extreme flood conditions.

Designers are cautioned that design and construction of buildings and structures in proximity to flood protective works will entail the review of all available maps, studies, and other information pertaining to the works, including the level of protection intended, the condition of the works, whether required maintenance has been performed, and any known drainage or ponding problems in the area. Designers are particularly cautioned that consideration of these aspects of risk associated with protective works is important when locating and designing critical and essential facilities that could be impaired if the protective works fail to provide the anticipated level of protection.

C1.4.3 Assignment of Flood Design Class to Buildings and Structures Table 1-1 of the standard defines Flood Design Classes 1 through 4. Each building and structure is to be assigned a Flood Design Class. The classes recognize the importance of occupancies, potential for damage, and threats to human life. Although similar, Flood Design Class is not equivalent to the occupancy category or risk category used by building codes and ASCE 7.

Most buildings and structures will be assigned to Flood Design Class 2. Buildings and structures in which a large number of persons may assemble typically are assigned Flood Design Class 3.

Flood Design Class 4 includes buildings that are essential facilities (see definitions) and buildings that provide services necessary for emergency response and recovery. This standard now requires Flood Design Class 4 buildings to be elevated or protected to at least the 500-year flood level.

Some owners, designers, and communities may elect to treat Flood Design Class 3 buildings as essential facilities and require them to meet the requirements for Flood Design Class 4 buildings. Designers should consider that any building or facility that is vital and needs to function during recovery after flood events should be assigned Flood Design Class 4.

For the purposes of this standard, only structures that represent a low hazard to human life and that pose little threat to other structures (during design flood conditions) should be classified as Flood Design Class 1. Table 1-1 footnote (a) indicates Flood Design Class 1 includes agricultural facilities as subject to the requirements of this standard, but certain agricultural facilities may be constructed with their lowest floors below the specified elevations, as allowed by the wet floodproofing provisions of Section 6.3. For example, FEMA recognizes that wet floodproofing may be appropriate for some agricultural structures—farm storage structures, grain bins, corn cribs, and general purpose barns—even though these structures are not used for those purposes specifically identified in the NFIP regulations for enclosures below the BFE (parking of vehicles, building access, storage) as being suitable for wet floodproofing. The NFIP will allow these types of agricultural structures to be wet floodproofed upon issuance of a variance from the community. FEMA Technical Bulletin 7, *Wet Floodproofing Requirements for Structures Located in Special Flood Hazard Areas* (FEMA 1993c) provides more information on this subject.

C1.5 BASIC DESIGN AND CONSTRUCTION REQUIREMENTS

C1.5.1 General It is essential that the designer understand the nature and magnitude of loads that will act on buildings and structures during design flood conditions. See ASCE 7 Chapter 5 and commentary (Standard Reference ASCE/SEI 2010). Furthermore, it is essential that appropriate load combinations be applied, especially for coastal areas (Coastal High Hazard Areas and Coastal A Zones). Design wind loads may act simultaneously with design flood loads in such areas. This is usually not the case in riverine areas; however, in riverine and lake floods of long duration where flood water depths and wide areas of water or low geography create long fetches, wind-driven waves may develop. In these areas, wave action may be important and the requirements of Chapter 4 should be considered reasonable for design and construction.

Although this standard is primarily intended for the design and construction of buildings, many of its provisions are also applicable to structures, such as gazebos, pavilions, bleachers, grandstands, towers, and similar structures that are not buildings. Such structures, if located in a flood hazard area, should be designed to resist flood loads associated with the design flood and should meet the applicable requirements of this standard.

Proper identification or calculation of design flood conditions is required before foundation type and other aspects of design can be considered. Design flood depths and velocities must be specified with a reasonable degree of accuracy; otherwise, flood forces computed during the design process will be unreliable. If flood hazard maps and studies do not specify flood elevations (e.g., as in the case of unnumbered A Zones) or flood discharge/velocity information, other sources of floodplain information should be consulted. Agencies listed in Section C1.3 should be contacted. In the absence of flood elevation and other data from FEMA or other agencies, professionals having the required expertise should be consulted.

Siting of buildings and structures in flood hazard areas involves land use and regulatory issues. The designer should consult the local jurisdiction and other agencies with zoning, land use, and regulatory authority to ascertain local and state siting restrictions or conditions. Many jurisdictions or state governing authorities may prohibit construction or other development activities in certain portions of the floodplain; many will impose specific requirements on siting, design, and construction. For example, some jurisdictions may require that new construction not cause any of the following to occur: (1) alter the flood path, (2) increase the flood discharge, (3) increase the flood velocity, (4) increase the flood elevation, (5) increase the area of flood inundation, (6) reduce flood storage, or (7) reduce available wetlands.

C1.5.2 Elevation Requirements The minimum required elevation of buildings and structures (lowest floor, dry floodproofing measures, flood damage-resistant materials, and location of utility equipment) is a function of two factors: the type of flood hazard area, and the Flood Design Class of the building or structure. Tables that specify the minimum elevations are found in Chapter 2 (flood hazard areas other than Coastal High Hazard Areas and Coastal A Zones), Chapter 4 (Coastal High Hazard Areas and Coastal A Zones), Chapter 5 (flood damage-resistant materials), Chapter 6 (floodproofing), and Chapter 7 (attendant utilities and equipment).

In general, elevations are higher in Coastal High Hazard Areas and Coastal A Zones, where the reference point on the building is the bottom of the lowest horizontal structural member of the lowest floor. In other flood hazard areas, the reference point is the top of the lowest floor system.

Flood Design Class 4 buildings and structures are essential facilities and warrant a higher level of protection than other structures. The minimum elevation tables specify that Class 4 structures are to be elevated to the BFE plus 2 or 3 ft, or to the DFE, or the 500-year flood elevation, whichever is higher.

Flood Design Class 3 buildings and structures that pose a high risk to the public or disruption to the community also warrant a high level of protection, and the elevation tables specify the minimum elevation to be BFE plus 1 or 2 ft, or the DFE, whichever is higher.

Flood Design Class 2, which includes most buildings and structures (including one- and two-family homes and townhouses), is specified to have the lowest floor located at a minimum elevation of BFE plus 1 ft, or the DFE, whichever is higher.

Flood Design Class 1 buildings and structures have the DFE as the minimum elevation requirement (or DFE plus 1 ft if dry floodproofed).

This standard requires buildings and structures to be elevated higher than the minimum requirements of the NFIP. The more stringent elevation requirement is based on post-disaster observations and insurance claims data that indicate buildings elevated to only the BFE are often subject to some degree of flood damage. The conclusion is that a relatively small increase in elevation not only reduces damage, thus facilitating more rapid recovery, but also reduces the cost of federal flood insurance.

C1.5.3 Foundation Requirements

C1.5.3.1 Geotechnical Considerations Foundation design should be based on an accurate identification of underlying soil and rock properties at the site of a proposed structure. Although not required by many floodplain management ordinances or other applicable construction standards, geotechnical investigations should be conducted at any construction site in a flood hazard area where available geotechnical data are insufficient for design purposes. Foundation design should take into account all potential impacts of soil saturation (especially during conditions of long duration flooding), consolidation, movement, expansion, and erosion, including the effects of long-term erosion that may occur at a site.

C1.5.3.2 Foundation Depth In addition to the requirements contained in Chapters 2 and 4 of this standard, the designer can refer to the body of literature dealing with flooding, flood-induced erosion, and local scour, for example, USACE (2006), Fowler (1993), and FHA (2012). The designer should consult professionals experienced in hydraulic and sediment transport analyses in the geographic area of interest.

C1.5.3.3 Foundation Walls and Wall Footings Use of foundation walls below the minimum elevations specified by Table 2-1 is permitted only where the walls and wall footings are designed and constructed to resist flood-related forces acting on those walls so as not to jeopardize the stability or integrity of the building or structure. In most instances, use of load bearing walls in flood hazard areas susceptible to high velocity flow, high velocity wave action, or other destructive flood forces will likely result in significant damage to or total destruction of a structure during design flood conditions. Load bearing foundation walls designed and constructed to withstand such loads may not be the most cost-effective solution, especially for construction less than three stories above the DFE.

Foundation walls that enclose areas below elevated buildings, also called solid perimeter walls or crawlspace foundations, must meet design specifications in Section 2.7 for flood openings. Solid perimeter wall foundations that are filled with engineered compacted fill that is topped with a slab (stem-wall foundations) are not subject to the specifications of Section 2.7.

Where foundation walls do not form enclosures, it is recommended that the walls be oriented parallel to the direction of flood flow. This will reduce hydrodynamic loads, debris impact loads, and other flood-related loads that will act during design flood conditions.

C1.5.3.4 Piers, Posts, Columns, or Piles In flood hazard areas other than Coastal High Hazard Areas and Coastal A Zones, use of piers, posts, columns, or piles, rather than placement of fill, will minimize reduction of flood storage and diversion of floodwaters and drainage onto other properties.

Failures of piers, posts, and columns have been observed following coastal flood events, especially in areas where designs have not adequately accounted for potential erosion and local scour. The designer is cautioned that the use of piles and columns may be more appropriate in coastal areas.

C1.5.4 Use of Fill In many areas of the country, fill is commonly used to meet the minimum elevation requirements in flood hazard areas, although structural fill is not allowed to elevate buildings and structures in Coastal High Hazard Areas and Coastal A Zones. In most cases, fill is placed, compacted, and shaped to receive a slab-on-grade. The placement of fill reduces flood storage and may increase local velocities; however, an engineering analysis of those effects typically is required only if the fill encroaches into the floodway. Check with the community regarding fill placement requirements.

The designer should examine other site conditions when determining the feasibility of the use of fill and the design of the fill. In addition to specifying the type of fill materials and the appropriate compaction specifications, designers should consider specifying fill slopes that are no steeper than 2 horizontal to 1 vertical to minimize erosion, protect the foundation, and facilitate maintenance. Vegetative cover that is resistant to flowing water can be specified for fill slopes subject to low velocities. Slopes subject to high velocities should be protected by other means, such as stone or riprap.

Designers should also be aware that the documentation of elevations required to demonstrate compliance (and for owners to obtain federal flood insurance) requires the determination of the elevation of the lowest adjacent grade. A structure that has the lowest adjacent grade above the BFE may be determined by FEMA to be out of the mapped special flood hazard area by the Letter of Map Amendment process. Consult the FEMA website or the community floodplain manager.

C1.5.5 Anchorage and Connections Designers should be aware that combinations of materials and connection types may be used and that possible failure modes associated with using those connection types should be considered. The continuous load path concept is important enough to emphasize that the load path is made through the connections used in the building frame. A failure in any of the possible modes potentially weakens the entire structure, depending on how the connections are loaded (shear, tension, compression, torsion, etc.). It is also important to point out that loads occur in two directions, vertical and lateral (horizontal), and load combinations must be applied in both directions.

Bolts are appropriate for wood-to-wood, wood-to-steel, and wood-to-concrete connections. Steel-to-steel and steel-to-concrete connections may be made using bolted connections or welded connections. Connections using bolts or welds should be designed using the appropriate load combination strategy and may be designed for tensile, compression, and shear loads. Connections should be made such that the full capacity of the connection can be developed. Bolt holes should be located with sufficient clearances to minimize wood crushing, concrete cracking, steel tearing, and other failure modes. Welding on plates of sufficient thickness should be conducted with welding electrodes of sufficient capacity to develop the full connection capacity.

C1.6 LOADS IN FLOOD HAZARD AREAS

Extensive commentary on design loads and load combinations is contained in Chapter 5 of ASCE 7 (Standard Reference, ASCE/SEI 2010).

C1.6.1 General Most loads acting on structures (such as wind loads) are typically on the order of tens of pounds per square foot, although some flood-related loads (loads from high velocity waters, wave action, debris impacts, and ice) may exceed those typical loads by a factor of 10 to 100, or more. Extreme loads such as these make design and construction impractical and cost-prohibitive for structures that are not elevated. Use of pile or column foundations allows waves and floodwaters to pass beneath elevated buildings and structures, thus minimizing the effects of those extreme loads.

C1.6.2 Combination of Loads Some flood loads must first be combined to determine the flood load F_a used in ASCE 7 Chapter 2, Combinations of Loads (ASCE/SEI 2010). For coastal areas, guidance on combining flood loads to yield F_a is found in FEMA P-55, *Coastal Construction Manual*, Section 8.5.12 (FEMA 2011).

CHAPTER C2

BASIC REQUIREMENTS FOR FLOOD HAZARD AREAS THAT ARE NOT IDENTIFIED AS COASTAL HIGH HAZARD AREAS AND COASTAL A ZONES

C2.1 SCOPE

This section defines basic requirements for design and construction of new construction and substantial improvement of existing structures located in flood hazard areas, including High Risk Flood Hazard Areas but not including Coastal High Hazard Areas (V Zones) and Coastal A Zones. Therefore, it applies in flood hazard areas commonly called "A Zones," which refer to flood zones labeled on FIRMs as Zone A, Zone AE, Zone AO, Zone AH, Zone A1-30, and Zone AR. Included are all mapped flood hazard areas along rivers, creeks, streams, lakes, and ponded areas, and some A Zone areas where coastal wave heights are less than 1.5 ft (see definition of Coastal A Zone).

Even if a site is mapped as an A Zone on a flood hazard map, it is strongly recommended that the designer verify the lack of high velocity wave action, high velocity flows, erosion, debris effects, ice jams, and other extreme flood hazards before applying the provisions of this chapter. See Chapter 3 for additional requirements for High Risk Flood Hazard Areas, and Chapter 4 for requirements related to Coastal High Hazard Areas and Coastal A Zones. See Chapters 5 through 9 for requirements related to flood resistant materials, dry floodproofing and wet floodproofing, attendant utilities and equipment, building access, and miscellaneous construction (decks and porches; concrete slabs; garages, carports, and accessory storage structures; chimneys and fireplaces; pools; and tanks).

C2.2 DEVELOPMENT IN FLOODWAYS

The floodway is generally considered to be the most hazardous portion of a riverine floodplain area—it conveys the greatest portion of the flood flow and is where floodwaters are more likely to be deep and have high flow velocity. Encroachment by construction, fill, or other development activities into the floodway can cause flood elevations to increase and can have other adverse consequences. Where feasible, designers and owners should locate new buildings and structures on sites that are outside of floodways. Some jurisdictions prohibit development, including new buildings, in floodways.

Construction, fill, and other development activities in floodways should not be permitted unless hydrologic and hydraulic analyses demonstrate the proposed activity will result in no decrease in effective conveyance and no increase in flood levels during the base flood. FEMA lists the numerical models it has determined to be acceptable to meet the agency's minimum requirements: http://www.fema.gov/national-flood-insurance-program-flood-hazard-mapping/numerical-models-meeting-minimum-requirement-0.

Many of the Flood Insurance Studies and FIRMs published by FEMA include regulatory floodways established by agreement between FEMA and the community. If FIRMs show floodways, new computations are generally not required, provided proposed development activities do not encroach into the designated floodway. However, lack of a designated regulatory floodway on a FIRM or Flood Hazard Boundary Map does not mean that protecting conveyance of the base flood discharge is not important—designers should check with the authority having jurisdiction before assuming flood hazards are uniform throughout the floodplain.

In instances where floodways have not been designated but where base flood elevations (BFEs) have been determined, Section 60.3(c)(10) of the National Flood Insurance Program (NFIP) regulations requires communities to prohibit new construction, substantial improvements, or other development (including fill), unless it is demonstrated that the cumulative effect of the proposed floodplain encroachments will not increase the water surface elevation of the base flood more than 1 ft. Some states and communities are even more restrictive—requiring demonstration that the cumulative effect of the proposed development will not cause any increase in the water surface elevation of the base flood. Some communities may require conveyance and/or storage compensation or other mitigation as a condition of permitting construction in the floodway.

Designers should check with the authority having jurisdiction to determine which development and construction activities are permitted in floodway areas and to determine any limitations, evaluations, or special permit conditions associated with those activities. The designer should be advised that some states and communities severely restrict construction and development in floodways and may prohibit residential construction entirely. For example, the state of Michigan prohibits all residential construction in floodways to ensure that the floodway remains uninhabited.

C2.3 ELEVATION REQUIREMENTS

A basic requirement for design and construction in flood hazard areas is that a structure should be capable of resisting flood forces and minimizing flood damage. This is usually accomplished by elevating the top of the lowest floor (including basement) to or above a specified elevation.

The lowest floor elevation requirements of this standard for Flood Design Class 2, 3, and 4 buildings and structures are more restrictive than NFIP requirements, resulting in reduction in damage and reduction in federal flood insurance premiums. Note also that elevation to the BFE provides no safety factor to account for uncertainties in flood frequency analyses and hydraulic modeling, nor do FEMA's BFEs account for changes in flooding because of watershed development, sea level change, changes in precipitation patterns, and other factors. This standard incorporates safety factors by requiring most buildings and structures to be elevated above the BFE (i.e., by including freeboard).

The designer is cautioned that elevation requirements included in Tables 2-1, 4-1, 5-1, 6-1, and 7-1 do not provide the same level of protection in all geographic areas. For example, 2 ft of additional elevation may be equivalent to a 200-year flood in one area and a 500-year flood in another. Designers may wish to investigate the issue thoroughly before accepting this standard's minimum elevation requirements—additional elevation might be desirable, especially for high value buildings and structures with critical and essential occupancies.

For buildings and structures governed by Chapter 2, floodproofing in accordance with Chapter 6 may be used in lieu of elevation. Designers are cautioned, however, that locating the lowest floor below the elevations required by Table 2-1 may result in increased construction and maintenance costs, may lead to significant damage if the level of dry floodproofing protection is exceeded by a larger flood, and may result in substantially higher flood insurance premiums.

Because of uncertainties in flood hazard mapping methodologies and because floods more severe than the base flood do occur, designers are advised to use the elevation and other requirements of this standard when designing buildings that are situated near, but not in, mapped flood hazard areas. In particular, at those locations, basement excavation and construction at a level below the flood elevation should be accomplished using appropriate design and construction.

It should be noted that the lowest floor of all new construction and all substantially improved buildings and structures is required to be at or above the elevations in Table 2-1, except for enclosed areas used solely for parking of vehicles, building access, and storage that are designed in accordance with this section, Flood Design Class 1 structures that are wet floodproofed in accordance with Section 6.3, and those areas permitted to be dry floodproofed in accordance with Section 6.2.

The standard permits basements in nonresidential structures but prohibits basements in residential structures. It is not the intent of this standard to prohibit residential basements in communities that have received basement exceptions from the NFIP. In accordance with Section 60.6(c) of the NFIP regulations, FEMA may allow communities to permit engineered basements beneath residential structures in very limited circumstances. These "basement exceptions" are granted on the basis of technical criteria requiring the community to demonstrate that the special flood hazard area is subject to shallow and low velocity flooding and that there is adequate flood warning time to ensure that all residents are notified of impending floods. In addition, communities subject to tidal flooding are not eligible for this NFIP basement exception. As of mid-2014, only 54 communities have received basement exceptions from the NFIP. The floodproofing provisions in Section 6.2 of this standard are appropriate for consideration when designing a dry floodproofed residential basement under the NFIP basement exception.

C2.4 USE OF FILL

C2.4.1 Structural Fill Structural fill should be used only in flood hazard areas not susceptible to high velocity wave action and other forces capable of eroding the fill. Structural fill should be suitable for its intended use and should not lead to unacceptable levels of expansion, consolidation, or movement during conditions of flooding. Structural fill should be granular and free-draining, wherever feasible. Structural fill used for foundation support and protection should be properly designed, constructed, and protected. Guidance on the protection of earth slopes to resist flood-related erosion from wave action or high flow velocities is available in several U.S. Army Corps of Engineers' publications, including USACE (2000), USACE (2004), and USACE (2006). For low flood velocities (5 ft/s, or less) adjacent to structural fills, fill and slope protection is normally achieved by creating slopes of less than 1 vertical to 1.5 horizontal and providing appropriate vegetative cover. Protection against moderate flood velocities (5–8 ft/s) will require even flatter slopes and/or the use of stone or other materials. Use and protection of structural fill in areas with flow velocities greater than 8 ft/s may not be feasible.

Designers planning placement of fill for the purpose of elevating buildings and structures and to remove sites and buildings from mapped special flood hazard areas are referred to FEMA Technical Bulletin 10, *Ensuring that Structures Built on Fill in or near Special Flood Hazard Areas Are Reasonably Safe from Flooding* (FEMA 2001b). Generally, excavation of basements into fill may expose the structure to residual risk and damage associated with flooding and saturated fill. Designers proposing basements in fill are advised that the elevation requirements of this standard apply to the basement, unless the building or structure is dry floodproofed in accordance with Section 6.2.

C2.5 SLABS-ON-GRADE AND FOOTINGS

Designers should consider extending structural fill that is used to support slabs-on-grade beyond the footprint of the foundation. Not only is access around the building improved, but the compacted fill that supports the building or structure is less likely to be disturbed during construction. The distance that the fill extends beyond the footprint of the foundation will vary as a function of the occupancy of the building and may be specified by communities.

C2.5.1 Use of Slabs-on-Grade Guidance related to reinforcement of slabs-on-grade can be found in ACI 302.1R, *Guide for Concrete Floor and Slab Construction* (ACI 2004).

Slabs constructed as floors for enclosed areas used for parking or storage should be constructed as specified in Section 9.3.

C2.7 ENCLOSURES BELOW THE DESIGN FLOOD ELEVATION

Enclosures below the DFE can be used only for parking of vehicles, building access, and storage, provided the requirements of this standard and the authority having jurisdiction are satisfied. The designer is advised to keep use of enclosures below the DFE to a minimum. The designer is also advised that enclosures below the DFE may result in substantially higher flood insurance premiums, and the insurance implications of enclosures should be investigated.

Designers should be aware of specific situations relative to crawlspace construction. To facilitate drainage of accumulated floodwaters, the interior grade of the crawlspace is to be at or above the exterior grade. (This standard defines a basement as any area of a building or structure that is below grade on all sides and basements are not permitted.) An exception to this is the situation that may be applicable in very limited circumstances: in flood hazard areas where the depth of water above grade (BFE minus ground elevation) is less than 2 ft, a below-grade crawlspace is allowed if the top of the footing is no more than 2 ft below grade (total wall height no more than 4 ft). Such below-grade crawl spaces are to be designed specifically in accordance with FEMA Technical Bulletin 11, *Crawlspace Construction for Buildings in Special Flood Hazard Areas* (FEMA 2001a).

C2.7.1 Required Openings in Foundation Walls and Walls of Enclosures The requirement for openings does not apply to nonresidential structures that are dry floodproofed in accordance with Section 6.2 of this standard.

Openings are required in foundation walls that enclose areas below the DFE in order to equalize hydrostatic pressures and prevent damage or collapse. Although openings in interior walls are not required by this section, designers should consider whether such openings are necessary to ensure that floodwaters can reach all portions of enclosed areas in order to minimize unbalanced hydrostatic loads on foundation walls and exterior walls. In addition, designers should consult with the authority having jurisdiction about requirements for such openings.

C2.7.1.1 Openings in Breakaway Walls The 2014 edition of this standard includes two significant changes to this section. First, the 2005 edition did not require openings in breakaway walls in Coastal High Hazard Areas, while this edition does; it requires openings in breakaway walls in all flood hazard areas. Second, the 2005 edition had two calculations for opening areas depending on flood zone and whether the walls were breakaway walls. This edition has a single calculation that is used for all openings, whether in nonbreakaway walls or breakaway walls.

Breakaway walls are intended to fail under wave loads. However, experience shows that some breakaway walls fail under water depths and wave conditions that are significantly less than base flood conditions. Experience also shows that openings in breakaway walls limit failure under these more frequent "shallow" flooding events. When breakaway walls stay intact during more frequent flooding, (1) the interiors of enclosed areas are not exposed to wind-driven rain and sand, (2) there is less debris added to floodwaters, and (3) building owners suffer less out-of-pocket costs to replace breakaway walls and repair flood damage.

C2.7.2 Design of Openings Flood openings can be engineered or non-engineered but should safely allow equalization of hydrostatic pressure outside and inside any enclosure below the DFE. FEMA Technical Bulletin 1, *Openings in Foundation Walls and Walls of Enclosures* (FEMA 2008d) has an expanded discussion of opening requirements.

C2.7.2.1 Non-Engineered Openings The prescriptive requirement for non-engineered openings specifies 1 sq in. of net opening area per square foot of enclosed area. The area of the enclosure is measured on the outside, which allows inspection and verification without requiring entry into the enclosed area or crawlspace. The sum of the net open area of all non-engineered openings provided for a specific enclosure must equal or exceed the required net open area for that enclosure. The minimum dimension of an opening in the wall must not be less than 3 in., as measured in any direction in the plane of the wall.

To determine the net open area of a non-engineered opening, the actual area through which water flows shall be measured. If a device is inserted into or affixed over an opening in the wall, the measurement of the net open area is not based on the dimensions of the opening in the wall but must account for the presence of the device. Area that is obstructed or covered in any way (other than by screening) shall not be counted as open area.

Any louvers, blades, screens, and faceplates or other covers and devices should be selected or specified so as to minimize the likelihood of blockage by small debris and sediment. Where experience has shown that a particular device or type of device has been blocked or clogged by flood debris or sediment, use of such devices should be avoided.

The requirement for flood openings may be satisfied by use of ventilation devices that commonly are used for foundation ventilation. When foundation ventilation devices are installed in accordance with Section 2.7.3, such devices are to be permanently disabled in the open position in order to satisfy the requirement that flood openings allow for the automatic entry and exit of floodwaters. Covers that must be removed and devices that must be manually opened do not satisfy the requirement for automatic entry and exit of floodwaters, and such covers and devices should not be specified by the designer.

Methods used by the ventilation industry to calculate the net open area of ventilation devices, for example, vents with fixed louvers, may be applied to determine the net open area for non-engineered flood openings. FEMA (2008d) advises that manufacturers of air vents typically indicate the number of square inches their air vents provide for air flow; that same number of square inches may be used for flood opening applications.

C2.7.2.2 Engineered Openings Engineered openings are intended to provide an alternate to the non-engineered or prescriptive requirement (1 sq in/sq ft of enclosed area), while still satisfying the required opening performance (automatic entry and exit of floodwaters with no more than 1 ft of difference between the exterior and interior water levels).

The NFIP and building codes require the design of engineered openings to be certified by a registered design professional. Certification requires more than simply applying the equation in this section; it requires consideration of a number of factors that represent expected base flood conditions. Worst case rates of rise and fall must be determined, opening shape and size (which affect flow efficiency) must be assessed, the potential for debris blockage must be evaluated, and the effects of any louvers, blades, screens, grilles, faceplates, or other covers and devices must be considered. The best means to certify performance is to test engineered opening devices under conditions that mimic a range of rising and falling floodwaters, preferably floodwaters that contain debris typical of floodwaters around buildings.

The standard indicates that a minimum rate of rise and fall of 5 ft/h should be used in the absence of reliable data or analysis. This rate of rise, only 1 in./min, is not representative of many flood hazard areas where both past experience or numerical modeling indicate much faster rates of rise and fall, suggesting designers should be cautious about relying on the 5 ft/h minimum rate. Depending on factors, such as watershed size, nature of land use, topography, and soil type, rainfall quickly runs off many watersheds, resulting in rates of rise and fall that exceed 1 in./min. Depending on coastal storm characteristics such as track, forward speed, and intensity, storm surge flooding also can result in rates of rise and fall that exceed 1 in./min. Information on rates of rise for specific areas may be available from stream gauges and tide gauges.

Information may also be available from sources familiar with past flooding characteristics at specific locations, including federal agencies such as USGS, NRCS, or NOAA; and state and local sources, such as floodplain management and building departments, emergency management agencies, public works departments, transportation agencies, and universities. Video documentation of rates of rise is becoming readily available for many flood events and may also serve as a basis for estimating rates of rise.

Unless an engineered opening is uniquely designed for a specific location, the performance of engineered openings should be tested over a range of rates of rise, including rates that are many times the minimum rate, so that designers have sufficient information on which to base decisions regarding whether to increase the total net area to account for faster rates of rise, as indicated in item 5 of this subsection.

Net open area of an engineered opening (A_o) should be measured in the same manner as for a non-engineered opening based on the actual area through which water flows, and excluding any area that is obstructed or covered (except by screening).

Discharge coefficients shown in Table 2-2 for unobstructed flow conditions were taken from or derived from standard hydraulics texts, and the values represent free-flowing, unobstructed openings. The coefficient represents the ratio of the actual flow through an opening divided by the ideal flow, where ideal flow is given by

$$Q = A_o(2\,g\,H)^{0.5}$$

where

Q = ideal flow (cfs)
A = net cross-sectional area of opening (sq ft)
H = depth of water above the bottom of the opening (ft)
g = gravitational constant (32.2 ft/s^2).

Note that the 2014 edition of this standard added a coefficient of discharge of 0.20 for openings that are partially obstructed. This standard classifies any opening with louvers, blades, screens, grilles, faceplates, or other covers or similar devices that will be in place during the design flood, as partially obstructed. A partially obstructed condition will also occur if debris blocks part of an opening even if there are no louvers, blades, screens, grilles, faceplates, or other covers or similar devices.

In the case of unobstructed circular openings, hydraulics references indicate the discharge coefficient for a circular, sharp-edged orifice in a vertical wall will be approximately 0.60 under low head conditions, which are required here (maximum head difference across opening equal to 1 ft). Coefficients for unobstructed rectangular and square openings were calculated for typical opening sizes (i.e., 12 in. × 12 in. or 8 in. × 16 in., corresponding to nominal masonry unit sizes) using the ideal flow discharge relationship as shown and the following discharge relationship for a contracted rectangular weir:

$$Q = 3.33(L - 0.2\,H)\,H^{1.5}$$

where

Q = flow through opening in cfs
L = horizontal length, ft
H = depth of water above the bottom of the opening, in ft.

Other unobstructed opening shapes were assigned a coefficient of 0.30 based on flow through a V-notch or trapezoidal weir.

If no louvers, blades, screens, grilles, faceplates or other covers, or similar devices are used, and if the designer is certain the potential for blockage by debris is low, unobstructed discharge coefficients in Table 2-2 between 0.25 and 0.60 may be used, depending on the opening shape.

This standard permits a more exact determination of a coefficient of discharge for a particular opening/device and debris condition by laboratory testing or numerical modeling of the opening/device, provided the resulting coefficient is no more than 10% greater than that shown in Table 2-2, and in no case shall a coefficient greater than 0.60 be used.

If an engineered device contains more than one open area through which water flows, the designer may assign a discharge coefficient to each open area based on the shape of each open area. This allowance is not intended to be applied such that portions of a partially obstructed device (see footnote b in Table 2-2) can be reclassified as unobstructed by making individual flow calculations for many small open areas in the device.

C2.7.3 Installation of Openings All openings, whether non-engineered or engineered, are to be installed as specified in this section. The minimum requirements for at least two openings and for openings in at least two walls are intended to minimize the likelihood that flood debris will block openings and prevent them from functioning as intended. Buildings on solid perimeter walls that are set into sloping sites may require installation of openings along the sloping grade, provided the openings are below the DFE (illustrated in FEMA 2008d).

Rising floodwater exerts unbalanced hydrostatic loads on a wall when the exterior water level rises above the interior floor level or when the interior water level is above the exterior grade. This is why the bottoms of openings are to be positioned relative to the higher of the final interior grade or floor and the finished exterior grade immediately under the openings. Designers are reminded that floodwaters of 1 ft depth may be trapped inside crawlspaces and enclosures, contributing to moisture problems during cleanup. To minimize this effect, openings may be positioned closer to grade.

Openings and opening devices may be distributed around the perimeter of an enclosed area and may be grouped. If grouped, whether side-by-side or stacked vertically, the installation requirements apply. If stacked vertically, a closely spaced group of openings or opening devices can be assumed to function as a single opening and the bottom of the lowest opening or opening device should be positioned to meet the height-above-grade requirement.

CHAPTER C3
HIGH RISK FLOOD HAZARD AREAS

C3.1 SCOPE

The nature of the hazards listed in Section 3.1 makes the identification of High Risk Flood Hazard Areas difficult and the design and construction in such areas problematic. In addition, the state or local authority having jurisdiction may have additional requirements or may not permit any building in these areas. The intensity, spatial extent, duration, and probabilities associated with these hazards are difficult to predict, leading to uncertainties associated with the delineation and management of High Risk Flood Hazard Areas. FEMA 116, *Reducing Losses in High Risk Flood Hazard Areas: A Guidebook for Local Officials* (FEMA 1987), provides general guidance for management of High Risk Flood Hazard Areas. Unfortunately, the scenic beauty of many areas in which high risk flood hazards occur attracts development interest and poses a serious challenge to floodplain managers and building officials.

The basic requirements of Chapter 2 apply in High Risk Flood Hazard Areas that are not identified as Coastal High Hazard Areas (V Zones) and Coastal A Zones, which are subject to the requirements in Chapter 4.

In general, Chapter 3 provides that new construction and substantial improvements shall not be allowed in many High Risk Flood Hazard Areas unless protective works are in place. However, the designer is cautioned that: (1) the same problems and uncertainties, as identified, that make the identification of High Risk Flood Hazard Areas difficult also make design and construction of protective works difficult; and (2) protective works, even if they function as designed, generally do not provide absolute or complete protection to an area. Whereas such works typically reduce the frequency of flooding, if a flood event exceeds the level used for design of the works, the resulting flooding may be very severe.

Construction of protective works (or reliance on existing protective works) to protect new buildings and structures in a High Risk Flood Hazard Area may be ill-advised, especially where uncertainties about flood hazards are great.

In High Risk Flood Hazard Areas, poor maintenance or improper operation of protective works and facilities, including pumping plants, can cause damage to or failure of those protective works and facilities and will almost certainly result in damage or destruction of structures behind the protective works. Therefore, to be considered as providing protection, all protective works should include implementation of a well-conceived plan for periodic inspection, maintenance, repair, and testing. In areas presumed to be protected by protective works that are in poor condition, designers should consider the effect of potential partial or complete failure, which would result in flood damage.

C3.2 ALLUVIAL FAN AREAS

Alluvial fan areas represent one of the most hazardous floodplain areas. Alluvial fans are geomorphic features characterized by cone- or fan-shaped deposits of boulders, gravel, sand, and fine sediments that have been eroded from upstream watersheds and then deposited on the adjacent valley floor. Alluvial fans can be found throughout the United States but are numerous in arid and semi-arid regions. Floods on alluvial fans have extremely unpredictable flow paths, so the design and construction standards address only design flood conditions, including the potential for surging, channel avulsion, and debris frontal waves.

For floodplain management purposes, alluvial fans are usually divided into areas of active alluvial fan flooding and areas of inactive alluvial fan flooding. Many alluvial fans can exhibit both active and inactive alluvial fan flooding hazards. Flooding that occurs on active alluvial fans is often characterized by debris and sediment-laden flows. Channel avulsion or overbank flows can result in unconfined flows on alluvial fans where flow paths are unpredictable and subject to lateral migration. In addition, these fast-moving flows present hazards associated with erosion, debris transport and deposition, and sediment transport and deposition.

FEMA uses three criteria to delineate active alluvial fan flood hazards: (1) flow path uncertainty below the hydrographic apex; (2) abrupt deposit and ensuing erosion of sediment as a stream or debris flow loses its competence to carry material eroded from a steeper, upstream source area; and (3) an environment where the combination of sediment availability, slope, and topography creates a hazardous condition for which elevation on fill will not reliably mitigate the risk. On FIRMs, flood hazard areas on alluvial fans are designated Zone AO (with or without depth number) or Zone AH (with BFE), and on some maps may be designated Zone A (without BFE).

Section 3.2 provides that new construction and substantial improvements are not permitted in three parts of alluvial fans: (1) at the apex of the fan; (2) in the fan's meandering flow path; and (3) in portions of the fan where there is evidence of active processes, such as braided channels, erratic flow paths and sediment transport. Active processes may also be characterized by one or more of the following: avulsion, deposition or aggradation, slope changes, distributary flow, scour/erosion, unconfined flooding, sheet flow, stream capture, lateral erosion, debris flow, and head cutting. Active areas are identified by careful review of the physical features of an individual fan, which are best obtained from topographic data, surficial geology maps, recent and historical aerial photographs, historical flood data, soils data, personal observation, and the experience and knowledge of the authority having jurisdiction.

In areas on alluvial fans where buildings and structures and substantial improvements are allowed, the requirements in Section 3.2 are in addition to requirements in Chapter 2. In designing foundations, designers are advised to set the lowest floor elevation higher than required and to be cautious when accounting for the potential for scour, especially if soils are relatively unconsolidated. In addition, consideration should be given to use of filled stem walls or columns or piers rather than perimeter walls (crawlspaces). Transported sediment may enter the crawlspace through flood openings, and large quantities of sediment inside a crawlspace not only limit access to underfloor utilities but could create unbalanced loads on perimeter walls that were not anticipated in the crawlspace wall design.

National Flood Insurance Program (NFIP) regulations in 44 CFR 60.3(c) apply to development on alluvial fans. In particular, Sections 60.3(c)(7) for Zone AO and 60.3(c)(11) for Zones AO and AH apply. See FEMA 165, *Alluvial Fans: Hazards and Management* (FEMA 1989) for a discussion of the NFIP regulations for alluvial fans.

In general, inactive alluvial fan flooding areas are not subject to active processes and are characterized by relatively stable flow paths. However, these areas can still be subject to flooding and may even exhibit some of the same characteristics as active alluvial fan flooding but not to the degree that causes flow path instability or uncertainty. Elevating buildings and structures in these inactive, lower risk areas of an alluvial fan will not ensure complete flood protection.

Near the fan apex, dry floodproofing measures are generally not recommended for individual nonresidential structures. Alluvial fan flood damage reduction potentially can be achieved by constructing debris collection dams near the fan apex with maintenance plans for the removal of debris after each flood. Flood conveyance channels would normally be designed to safely transport high velocity discharge from the debris collection dam and downstream tributaries to a disposal location such as a large river. For such facilities intended to protect FEMA-mapped alluvial fan areas, see FEMA guidance in FEMA (1989) and NFIP regulations (44 CFR 65.13). The U.S. Army Corps of Engineers flood protection project for the City of Los Angeles is an example of alluvial fan flood damage reduction.

C3.2.1 Protective Works in Active Alluvial Fan Areas Areas on active alluvial fans identified in Section 3.2 where new construction is not permitted can be modified by protective works. This section provides performance expectations for such protective works. For protective works intended to protect FEMA-mapped alluvial fan areas, see FEMA guidance and NFIP regulations (44 CFR 65.13).

Extensive damage experienced on alluvial fans generally results from floods that exceed the design parameters of protective works and flood control structures. Damage can be particularly extensive to structures that are located directly in the flow path.

Alluvial fan flood mitigation can take two approaches: upstream storage of floodwaters, debris, and sediment, or conveyance of flood flows off the developed fan. Communities may wish to reserve a storage area immediately down-fan from the apex. If existing buildings and structures occupy the reserved storage area, communities may want to consider measures to not allow reconstruction after such structures incur substantial damage. Alluvial fan communities may also choose to restrict development in areas reserved to convey flows. However, unless a whole alluvial fan flood damage reduction project is constructed, simply reserving flood conveyance will not ensure containment of the design flood. Some combination of the two approaches may be considered. Flood hazard avoidance on alluvial fans is strongly encouraged. In some cases, the local authority should consider restriction of development on the entire alluvial fan.

Designs for channels or levees intended to contain and convey mud and debris flows must consider the potential for sediment deposition in the channel that reduces the conveyance capacity. Freeboard in excess of 5 ft should be considered for watersheds with large boulders and large volumes of sediment. Impact pressure on walls during a mudslide (i.e., mudflow) event should consider the largest diameter boulder found on the alluvial fan traveling at a velocity equal to or in excess of the peak discharge velocity of floodwaters.

C3.3 FLASH FLOOD AREAS

Flash flood areas present a serious risk to human life. Because of this high risk, construction of buildings and structures in flash flood areas should be avoided. Good examples that illustrate how dangerous flash flooding can be are the Big Thompson River downstream of Estes Park, Colorado (1976 and 2013) and Gatlinburg, Tennessee (1973 and 1994). Both of these locations are characterized by steeply sloped mountain streams with limited channel capacity and are subject to high intensity, short duration storms. Although flood warning systems have been developed to alleviate the threat to life in many areas subject to flash flooding, such systems cannot completely eliminate the threat because of the quickness of the flood events.

FIRMs do not identify which special flood hazard areas are subject to flash flooding. Flood Insurance Studies may contain some information that can be used to estimate which waterways may be subject to rapid onset flooding. Because of the nature of analyses of rainfall and runoff data with respect to identifying flash flood areas, designers are cautioned that such analyses should be conducted by qualified professionals. In any given region, community officials, state geological agencies, U.S. Geological Survey, U.S. Army Corps of Engineers, or the Natural Resources Conservation Service, may have knowledge of watersheds that have experienced flash flooding that can be used to extrapolate to similar watersheds.

C3.3.1 Protective Works in Flash Flood Areas Protective works in flash flood areas should provide protection against floods more severe than the 100-year flood to significantly reduce the threat to life. Rather than site-specific protective works, flood protection in these areas typically involves flood control reservoirs, which may be hard to justify because of their large cost, potential environmental impacts, and reservoir regulation difficulties.

C3.4 MUDSLIDE AREAS

The term "mudslide" as defined in this standard is consistent with the NFIP regulations. It is intended to include only the classes of flow that are technically recognized as mudflow and mud flood and not include landslides.

Mudflows are nonhomogeneous, non-Newtonian, viscous, transient hyperconcentrated sediment flood events whose fluid properties change dramatically as the flow progresses downslope. Viscous mudflow behavior is a function of the fluid matrix of water and fine sediments with a significant yield stress that must be exceeded to initiate motion. Mudflow areas are identified by irregular deposits, poorly sorted boulders, debris piles, natural levees, and large boulders transported long distances on mild slopes. Alluvial fan sediment deposits have a significant percentage of slits and clays. Damage to structures is caused primarily from frontal wave impact, mudflow deposition and inundation, boulder impact, and lateral loading.

Mud floods are hyperconcentrated sediment flow, turbulent in nature, with essentially water flow behavior. Mud floods have no yield stress, and large sediment particles will settle in a quiescent flow condition. Mud flood areas are identified by eroded channels, small levees, sorted boulders, and sediment deposits with very few fine sediments (silts and clays). Mud floods will cause damage to structures primarily from frontal wave impact, inundation, boulder impact, and lateral loading.

Mudflows are hyperconcentrated flows with a sediment concentration ranging from 45% to 55% by volume. Inundation and deposits of mud, impact of mud frontal waves, flow competence (ability to transport large boulders), and high lateral loadings during mudflows can cause structures to collapse or be moved off their foundations.

Landslides, which have sediment concentrations that exceed 55% by volume, can occur when hillsides become saturated from runoff infiltration or increased groundwater levels and are not considered to be flood related by this standard. Slope stability analyses are required to identify potential landslide areas. In earthquake-prone areas liquefaction should be considered in the hazard risk assessment.

Many factors can contribute to the initiation of mudslides or landslides, including vegetation removal (including removal by wildfire), changes in soil moisture conditions, removal of the toe of the slope, or increased loading. Usually, areas of potential landslides have surface evidence of this potential or there is historical information where landslides have occurred in the vicinity.

FIRMs do not identify which special flood hazard areas are subject to mudslides. Because of the nature and complexity of analyses of mudslides, designers are cautioned that such analyses should be conducted by qualified professionals. In any given region, community officials, state geological agencies, U.S. Geological Survey, U.S. Army Corps of Engineers, or the Natural Resources Conservation Service may have knowledge of areas that have experienced mudslides that can be used to extrapolate to similar areas.

C3.5 EROSION-PRONE AREAS

Erosion is the physical process through which coastal, lake, and river shorelines retreat or are undercut. It also refers to the general wearing away of the ground, which lowers elevations over an area. In this standard, erosion is distinct from "local scour," which refers to the localized removal of soil by moving water around foundation elements.

Historical aerial photographs, topographic maps, nautical charts, and survey data can be used to investigate erosion and estimate erosion rates. Many states already carry out such investigations and have established construction setbacks based on long-term average annual erosion rates, especially in coastal areas. Some states have identified coastal areas subject to erosion up to and during 1% chance (100-year) storms. Especially in regions with unconsolidated soils where river channels actively migrate, some states and communities may have identified river shorelines that are subject to erosion or may have established setbacks.

FIRMs do not identify specific areas subject to erosion. Designers should research state and local erosion zone and shoreline setback requirements. Because of the nature and complexity of analyses of flood-induced erosion and long-term erosion rates, designers are cautioned that such analyses should be conducted by qualified professionals.

C3.6 HIGH VELOCITY FLOW AREAS

This standard defines high velocity flow as water moving adjacent to buildings and structures that has velocities greater than 10 ft/s.

FIRMs do not identify which special flood hazard areas are subject to high velocity flows. Flood insurance studies may contain some information that can be used to estimate velocities, especially if floodways have been designated (see floodway data tables). High velocity flow areas can be identified based on site location, historical observation of flood conditions at the site, and hydraulic analyses.

Limiting buildings and structures in high velocity flow areas is based on the potential for severe scour unless protection measures are provided to divert high velocity flows or otherwise protect the elevated structure (FHA 2012). High velocity flows exhibit significant turbulence and may have standing waves. The engineering models that are used to compute water surface elevation also compute the energy grade line, which is, among other things, an approximation of the height to which waves and turbulence may rise. The energy grade lines may be considerably higher than the water surface elevation. If structures are to be constructed in high velocity flow areas, designers are advised to examine available hydraulic models to determine the computed energy grade line and should use this to represent depth in calculating flood loads.

C3.8 ICE JAM AND DEBRIS AREAS

Ice Engineering (USACE 2002) provides additional guidance for design in ice jam and ice flow areas.

FIRMs do not identify which special flood hazard areas are subject to ice jams and debris hazards. Community officials, state geological agencies, U.S. Geological Survey, U.S. Army Corps of Engineers, Natural Resources Conservation Service, and state departments of transportation may have information on waterways that have experienced ice jam and debris hazards.

One way to identify whether a flood hazard area at a specific location is subject to ice jam and debris hazards is to consider the flood depth and velocity. If upstream sources of ice or debris are present and if the product of the flood depth and velocity at the site exceeds some critical value, say 4 ft^2/s, then potential ice jam and debris hazards exist. The rationale for the depth-velocity criteria is based on the concept that small objects transported at velocities up to 4 ft/s, or larger objects transported at lower velocities, would not be expected to cause significant damage to a structure or its foundation. However, the designer should consider this analysis as preliminary and conduct a more detailed analysis at a given site based on site conditions and structure design. The designer should also be careful to avoid a design that traps floating ice and debris at the structure, thereby imposing potentially large and unaccounted for loads.

ASCE 7 Section 5.4.5 and commentary (Standard Reference, ASCE/SEI 2010) provides guidance on impact loads.

CHAPTER C4
COASTAL HIGH HAZARD AREAS AND COASTAL A ZONES

C4.1 SCOPE

This chapter defines the requirements for buildings and structures that are subject to damaging wave action in Coastal High Hazard Areas (V Zones) and Coastal A Zones. Coastal High Hazard Areas and Coastal A Zones are areas where wave forces can be significant and must be considered in design for (1) areas where base flood conditions and water depths are sufficient to support breaking waves equal to or greater than 3 ft in height, (2) areas that are subject to wave runup depths of 3 ft or greater during the base flood, and (3) areas subject to base flood wave heights between 1.5 and 3 ft (Coastal A Zones). The rationale for selecting the 3 ft wave criterion is described in the U.S. Army Corps of Engineers' *Guidelines for Identifying Coastal High Hazard Zones* (USACE 1975).

Previous editions of this standard required designers to determine whether Coastal A Zone wave conditions existed and to treat the Coastal A Zone like Coastal High Hazard Areas. Starting with this edition, Coastal A Zones are treated like Coastal High Hazard Areas only if a Limit of Moderate Wave Action (LiMWA), as defined in the standard, is delineated on the FIRM, or if the authority having jurisdiction designates areas as Coastal A Zones.

Designers are cautioned that waves less than 3 ft in height can cause damage to structures and should be considered in load calculations and in foundation designs, whether or not a Limit of Moderate Wave Action has been delineated.

C4.1.1 Identification of Coastal High Hazard Areas and Coastal A Zones The wave hazard classification employed by this standard is slightly different from that used by the National Flood Insurance Program (NFIP) in mapping flood hazard areas. NFIP floodplain mapping procedures restrict consideration of wave effects to Coastal High Hazard Areas (i.e., the V Zone), although wind-driven wave effects may also be important in some lake and wide riverine areas. This standard intends for structures subject to damaging wave forces to be designed and constructed to resist those forces, regardless of location or the nature of the flooding. Hence, this standard is more restrictive than the NFIP regulations with respect to the identification of flood hazard areas subject to damaging wave action.

Designing a structure to withstand wave forces or other extreme forces can result in significant structural, architectural, and site planning modifications, as well as an increased cost of construction. Conversely, failure to design for these loads when they occur may result in total or near-total loss of the structure and its contents.

Proper application of this standard may result in designing and constructing for high velocity wave forces even though a site has been designated as an A Zone by the NFIP. Designers using this standard should verify if high velocity wave forces are a significant design issue at a site before following the requirements of Chapter 2 or Chapter 4. Reliance solely on the V Zone and A Zone designations made by the NFIP may not be sufficient, especially if the FIRM is old or if physical or hydraulic conditions have changed since the NFIP flood study was completed; it is possible that a structure located in an area designated by the NFIP as an A Zone will be subject to damaging wave forces.

In order for a site to be designated "subject to high velocity wave action" by the NFIP (i.e., V Zone), one of the following conditions must hold:

1. Water depths at the site must be sufficient to support 3 ft wave heights, wind or seismic forces must be capable of generating 3 ft wave heights during the base flood, and local conditions must be such that the 3 ft waves can propagate over the surface of the flood waters and reach the site (see Fig. C4-1);
2. Wave runup depths during the base flood must equal or exceed 3 ft (see Fig. C4-2);
3. Wave overtopping of a seawall, revetment, or other barrier must exceed a threshold value established by FEMA coastal mapping guidelines; or
4. A "primary frontal dune" must be present and mapped by FEMA as Zone V.

In order for an area to be designated as a Coastal A Zone, either FEMA must delineate the LiMWA on the FIRM or the community must designate certain areas as Coastal A Zone. Coastal A Zones are areas where waves between 1.5 ft and 3 ft occur. Commentary Section C 1.2 Definitions explains the terms.

A designer can determine whether or not water depths are sufficient to support 3 ft wave heights (high velocity wave action condition 1, as mentioned) using a common approximation that states the minimum stillwater depth capable of supporting a 3 ft wave will be 1.28 times the breaking wave height (see NAS 1977). Using this approximation, a 3 ft wave will not occur in a stillwater depth less than approximately 3.8 ft.

Stillwater depths at a site are assessed; calculations should be based on the stillwater flood elevations minus the eroded ground elevation expected at the site during the design flood event, not the preflood ground elevation at the site. Thus, a designer should obtain information on expected design storm and long-term erosion or calculate the erosive effects of the design flood on local soils, taking into consideration the generalized erosion that may result from water and waves, and any localized scour because of the interaction of water and waves with the structure being designed.

Any time stillwater flood depths exceed 3.8 ft, the *potential* for high velocity wave action exists. However, the presence of a 3.8 ft stillwater flood depth at a structure site is not, by itself, sufficient to classify the area as subject to high velocity wave action. In order for that designation to be accurate, the second condition stated previously must also hold. In other words, wind

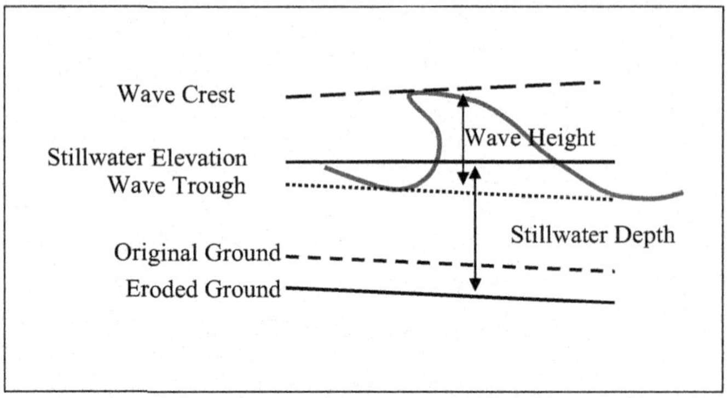

Fig. C4-1. Definition sketch for stillwater depth and wave height.

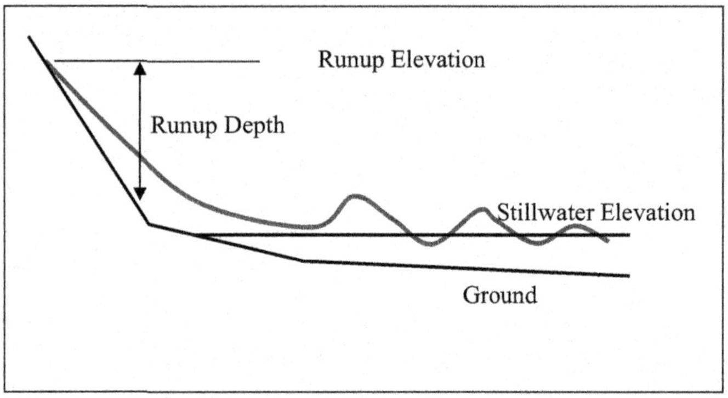

Fig. C4-2. Definition sketch for wave runup.

or seismic forces must be sufficient to generate waves equal to or greater than 3 ft in height at the shoreline, and water depths and obstructions between the shoreline and the site of interest must not reduce wave heights below 3 ft at the site.

In the case of wind-generated waves (the most common damaging waves), well-established procedures exist for quantifying the generation, propagation, and dissipation of waves (see NAS 1977, USACE 2006). These references will allow a designer to calculate expected wave heights and estimate wave runup and overtopping at the site of interest, thus allowing the designer to determine whether a structure will be subject to high velocity wave action.

If a Coastal A Zone is delineated or designated, this standard applies the elevation, foundation, and other requirements that previously were applied only to areas subject to high velocity wave action (V Zones). The justification for increasing the design and construction requirements in Coastal A Zones is based, in part, on recent post-storm damage investigations, a critical review of the original establishment of the 3 ft breaking wave as the landward limit of the Coastal High Hazard Area, and expert opinion solicited by FEMA. The justification and data that support the expansion are addressed in Jones et al. (2001) and summarized as follows:

1. The characteristics of flood hazards in V Zones and Coastal A Zones are largely the same—wave action, high velocity flow, erosion, and floating debris. Only the magnitudes of these hazard characteristics vary, with little diminution in magnitude in the portion of the flood hazard area that is adjacent to and inland of the V Zone. In other words, base flood conditions in Coastal A Zones are more like V Zone conditions than conditions in areas subject principally to inundation (many A Zones).
2. In 1977, the U.S. Army Corps of Engineers (which provided much of the justification for the NFIP's original use of the 3 ft wave height) conducted an analytical study and determined that a 2.1 ft breaking wave height represented the wave height threshold for damage to conventional wood-frame walls. A more recent full-scale laboratory study, cited in Jones et al. (2001), determined that 1.5 ft breaking waves consistently caused failure of the bottom plate of wood-frame walls.
3. The NFIP's A Zone depth-damage functions are based largely on many thousands of flood insurance claims paid on buildings located in flood zones subject largely to inundation (without waves). The V Zone depth-damage functions are based largely on claims paid on buildings located in V Zones but supplemented by expert opinion solicited by FEMA. (There are far fewer V Zone claims than A Zone inundation claims.)
4. Prior to about 2000, FEMA had no depth-damage functions for Coastal A Zones. Coastal A Zone depth-damage functions were subsequently developed based on analysis of Hurricane Opal damage in 1995 to structures at Pensacola Beach, Florida, and supplemented later by expert opinion. The Hurricane Opal analysis found that application of A Zone depth-damage functions to structures subject to wave height less than 3 ft resulted in significant under-prediction of actual damage, but application of V Zone damage functions to those same structures resulted in a good approximation of damage.

C4.2 GENERAL

Specific guidance for calculation of wave loads is found in ASCE 7 (Standard Reference, ASCE/SEI 2010).

Designing for wave forces requires an understanding of the mechanisms by which waves can affect structures. If waves are of sufficient height, they can cause damage in the following ways:

1. Waves may break against the side of a structure, causing horizontal loads at least an order of magnitude higher than wind loads;
2. Waves may introduce significant drag, inertia, and other forces on foundation and other structural members supporting elevated structures;
3. Waves may "peak up" or be redirected beneath an elevated structure and create large hydrodynamic uplift forces;
4. Waves may break near the structure and run up or over nearby terrain or erosion control structures, striking the structure;
5. Waves may cause significant erosion of supporting fill beneath non-elevated structures or by causing significant local scour around foundation elements supporting elevated structures; and
6. Waves can convey debris at high velocity, causing impact damage to buildings and supporting members.

In Coastal High Hazard Areas and Coastal A Zones where the aforementioned wave effects are likely to impact structures, designers should take the anticipated wave loads and other loads during design flood conditions into consideration, especially for the design of the foundation and structural supports. Guidance for design and construction of elevated buildings and structures may be found in FEMA P-347, *Above the Flood: Elevating Your Floodprone House* (FEMA 2000) and FEMA P-55, *Coastal Construction Manual* (FEMA 2011). Designers are advised that the authority having jurisdiction will require documentation of the design to be signed and sealed by the designer (commonly referred to as a V Zone certification).

C4.3 SITING

Earlier editions of this standard required buildings in Coastal High Hazard Areas and Coastal A Zones that were to be substantially improved (including buildings repaired after incurring substantial damage) to be located landward of the reach of mean high tide. This requirement effectively prevented existing V Zone and Coastal A Zone buildings (many on piers and over water) from being substantially improved or repaired without being relocated landward of the mean high tide line. In some cases these buildings can be made to comply with all flood resistant design and construction requirements except the siting requirement. This edition removes the landward of mean high tide siting requirement for substantially improved and substantially damaged buildings in V Zone and Coastal A Zone but leaves the siting requirement in place for other new construction. This edition is now consistent with NFIP minimum requirements that prohibit new construction over water in V Zone.

Buildings located seaward of mean high tide and subject to wave action are in especially hazardous locations, and their design is challenging. The justification for locating new buildings landward of the reach of mean high tide is well founded, given the reduction in flood hazards and flood forces as one moves landward from the shoreline. Locating buildings inland from the shoreline is consistent with the intention of this standard to promote public safety and protect property. Thus, designers (and owners) should consider relocating substantially improved and substantially damaged buildings in Zone V and Coastal A Zones to sites that are landward of mean high tide.

This edition of the standard recognizes, however, that some coastal communities have existing buildings on piers or wharves over water, where those piers or wharves were designed to resist the forces associated with coastal flood events. Many of these buildings were originally port facilities or water-dependent uses and have since been converted to other uses. This edition of the standard permits such buildings to be substantially improved or repaired without relocation, provided they can be brought into compliance with the foundation, elevation, and other prescriptive and performance requirements of this standard.

Mangrove stands and sand dunes are natural barriers that reduce the landward transmission of waves and high velocity flows. The weakening or removal of mangrove stands or sand dunes as a result of construction or development activity can adversely impact properties by allowing waves and high velocity flows to penetrate further landward, exposing lands and structures to increased wave and hydrodynamic forces. No construction or development should take place that leads to a reduction of the wave and energy dissipation characteristics of sand dunes or mangrove stands.

In areas protected by sand dunes, special care must be taken during the planning and construction or development to prevent degradation of the integrity of the dune system. Artificial alteration of dunes in a Coastal High Hazard Area that can increase potential flood damage is prohibited by the NFIP. Alteration of these features in Coastal A Zones (although not prohibited by the NFIP) may also result in increased damage. Excavation of dune material should be prohibited, or, where temporary disturbance is necessary, replacement of disturbed material should be required. With the exception of dune crossovers (access ways), construction and development activities should not take place on a dune.

Extreme care should be taken in siting buildings and structures in coastal areas, in recognition of the dynamic nature of many open coast, inlet, and bay or lake shorelines.

Sites located near tidal inlets can experience higher than anticipated BFEs and wave heights because of wave refraction and wave-current interactions. As a result, when locating a building or structure near a tidal inlet, it is prudent to add an additional foot (or more) of elevation (freeboard) as a factor of safety against these local wave effects that may not be captured adequately by flood models and flood hazard maps. Similarly, designers should be aware of the dynamic nature of shorelines in the vicinity of tidal inlets and attempt to identify trends or patterns in shoreline movement (using historical data) that should be considered in siting buildings and structures.

Along many coasts, average annual long-term horizontal erosion has been documented at rates of 2 to 5 ft/year, or more; in rare instances, rates exceeding 20 ft/year have been reported. Construction setback lines, based on long-term shoreline change rates, have been established by many state and local governments. Designers should know that short-term erosion rates can greatly exceed long-term rates and that construction landward of an established erosion setback line is no guarantee that the structure will not be affected or undermined by erosion. Therefore, the designer should be aware of established setback regulations and seek published long-term and short-term erosion data based on historical shoreline changes. State coastal zone management agencies typically maintain or have access to historic shoreline data and should be consulted.

C4.4 ELEVATION REQUIREMENTS

The elevation of any enclosed area used for purposes other than parking of vehicles, building access, and storage must be at or above the elevation of the lowest horizontal structural member, as established by this standard. The NFIP requires that the bottom of the lowest horizontal structural member be set at or above the BFE, whereas this standard has chosen to be more restrictive, requiring varying amounts of elevation above the BFE, depending on the Flood Design Class assigned to a building (see commentary for Section 1.4).

Previous editions of this standard specified elevation as a function of orientation of the lowest horizontal structural member relative to the direction of wave approach. However, at many sites it is difficult to determine which direction is the most likely direction of wave approach. Therefore, this edition eliminates that factor and simplifies decisions by both designers and local officials (see Table 4-1).

C4.5 FOUNDATION REQUIREMENTS

C4.5.1 General Designers should be aware that foundations in areas subject to wave action are often exposed to erosion, both storm-induced and erosion associated with long-term recession of the shoreline, and local scour around foundation elements. Thus, design of foundations must account for loss of soil associated with multiple factors. In some cases, flood loads will be increased with soil lowering. Soil loss around vertical foundation elements will increase the unbraced length and may require more embedment, bracing, or larger piles or some combination of all of these design modifications.

In mid-rise buildings and structures greater than about four stories, wind or seismic loads may be a controlling factor in the design of the structural frame, but flooding may play a significant role in the design of the lower portion of the structure. In buildings and structures of three or four stories, the coincident effects of both winds and potential wave action will be major considerations. Although seismic loads may govern design, the probability of simultaneous occurrence of earthquakes and coastal flooding is very low. The exception to this would be seismic events that trigger tsunamis. This standard does not require foundations to be designed to resist flood loads associated with extreme tsunamis, such as those that may be used for tsunami evacuation planning.

For taller buildings and structures, wind and seismic forces become increasingly greater and usually determine the governing design parameters for the structure. Foundation and framing systems must be able to resist the vertical forces from dead and live loads, as well as lateral forces from wind and flooding. Although shear walls may be appropriate for use in mid-rise and taller structures, they must be properly engineered to resist all load combinations.

The free-of-obstruction requirement is intended to permit free passage of floodwaters under elevated buildings and structures and through aboveground building foundations. Guidance on this requirement, and on potential obstructions that are part of the structure and that exist on the site, is in FEMA Technical Bulletin 5, *Free-of-Obstruction Requirements* (FEMA 2008c). FEMA Technical Bulletin 4, *Elevator Installation* (FEMA 2010a) provides additional guidance on elevator installations in V Zones.

In the absence of more detailed analysis, it is recommended that the flood loads in Coastal High Hazard Areas and in Coastal A Zones be taken as the greater of breaking wave loads or hydrodynamic loads on the row of columns closest to the flooding source and the hydrodynamic load on all other columns supporting the structure; see FEMA P-55, *Coastal Construction Manual* (FEMA 2011). This loading condition accounts for broken waves landward of the first row of columns. In addition, flood-borne debris can occur and should be accounted for as a point load on some critical column that is necessary for building support to prevent partial or complete collapse should the column be damaged and the load carrying capacity severely diminished.

C4.5.1.1 Shallow Foundations in Coastal High Hazard Areas Where shallow foundations must be used because of soil conditions, including the presence of rock, columns may be used with supporting footing, mat or raft foundations, provided they have sufficient strength to resist anticipated combination of flood loads, including hydrostatic, hydrodynamic, wave, and debris loads.

C4.5.1.2 Shallow Foundations in Coastal A Zones While spread footing, mat and raft foundations are permitted by the NFIP in A Zones (including Coastal A Zones), the wave and flow conditions expected during the base flood may be sufficient to cause undermining and significant damage to buildings and structures supported by those foundations. Designers should verify design flood conditions and make sure that any shallow foundations can satisfy the requirements of this section; otherwise, deep foundations should be used.

C4.5.3 Foundation Depth Foundations for buildings and structures in areas subject to wave action should be designed and constructed in recognition of the dynamic nature of these areas. In addition to wave forces and flood-induced erosion, many areas are subject to long-term or cyclic erosion that should also be considered. This standard requires consideration of erosion and scour as part of foundation design. Thus, the designer should first obtain or calculate the expected grade elevation after wave- or current-induced general erosion at the site and then consider the effects of local scour.

C4.5.4 Use of Fill It is not the intent of this standard to prevent the construction, maintenance, or reconstruction of sand dunes, or to prevent placement of minor amounts of nonstructural fill for landscaping, site grading, or drainage purposes.

C4.5.5 Deep Foundations Deep foundation embedment should be calculated based on the applied loads (see Section 1.6) and geotechnical conditions (see Section 1.5.3). Many pile foundations, particularly for small residential buildings, have been installed in accordance only with past practice and not by application of a recognized design procedure that accounts for both vertical and horizontal loads while also taking scour and erosion into account. The result has been undermining and foundation failure in many instances.

If soil conditions encountered during construction or installation are different from those used to calculate pile embedment ("unexpected" conditions), then embedment and foundation calculations should be checked for validity, and the design modified if necessary.

Section 4.5.5 specifies that pile foundations should be embedded to no shallower than 10 ft below mean water level (MWL), or to an elevation higher than −10 ft MWL if tied into a nonerodible soil or rock strata, or if foundation calculations demonstrate embedment to a shallower depth provides all required structural support and stability. If soil conditions prevent pile tips from reaching −10 ft MWL, piles should be embedded to refusal. The recommended −10 ft MWL embedment depth is a prescriptive minimum and should provide for adequate building support under typical erosion and scour conditions. However, some post-storm observations indicate this minimum embedment depth was not sufficient in some locations. Also, refusal may not be a sufficient depth criterion if erosion can occur to the depth of refusal strata; in such situations, foundation members should be anchored to refusal strata. The designer must be satisfied that there is not a failure mode for the pile that can occur when embedded less than required for support by frictional resistance. A connection of the pile base to the supporting strata must be sufficient to resist the lateral loads imposed by wind, seismic, or flood events and the appropriate load combinations.

Refusal should be defined by the geotechnical investigation but can be evaluated in the field by the number of blows per inch of pile penetration. For example, in the case of wood piles—solely for the purpose of determining embedment depth and not for meeting bearing requirements—refusal can be taken to mean 4 to 5 blows per in. if driven using a standard 5,000-lb hammer dropped from a minimum 3 ft lift, or using an equivalent hammer providing a minimum of 15,000 ft-lb of energy. For prestressed concrete piles, 6 to 8 blows per in. would be a reasonable value, and for steel piles 12 to 15 blows per in. (Bowles 2001). Where the final pile depth is less than that called for by the design, a verification is needed to determine if the "as-built" foundation will still resist the anticipated loads during periods of erosion and scour.

Foundation systems not covered specifically in this standard (e.g., helical piles or micropiles) may be considered for certain conditions and installation situations, such as low head room or minimal working space. If these foundation systems meet the load and other requirements of this standard, then the pile manufacturer's installation instructions must be strictly adhered to.

C4.5.5.4 Wood Piles Specification of minimum pile diameters is based on experience. For example, construction in coastal North Carolina has used 8 × 8 sq in. piles almost exclusively since the late 1970s, and the state building code has required them since the early 1980s. Field investigations of pile-supported residential structures after hurricanes since the 1990s have found few apparent problems with use of 8 × 8s in North Carolina, as long as erosion around the foundation was not significant. However, those post-storm observations indicated that as erosion increased and/or the number of stories increased, actual or incipient foundation failures became more frequent. This suggests there are undoubtedly pile lengths and design load combinations where larger piles become appropriate.

In 2008 Hurricane Ike created very large, deep scour holes around wood pile foundations in coastal Texas, exposing sufficient pile length either to cause the piles to fail in shear or to allow the piles to rotate at the bottom owing to insufficient embedment. Soil conditions can affect the amount of erosion and scour, and these conditions can be very localized. Knowledge of local soil conditions is important to properly designing a deep foundation.

The designer is cautioned that square timber piles do not take treatment well where heartwood is exposed at the face. In addition, scour is known to be greater surrounding square shapes than rounded shapes.

C4.5.5.6 Concrete-filled Steel Pipe Piles and Shells A minimum material thickness for steel pipe piles, shells, and casings in excess of that required by service conditions may be required to resist high collapse pressures at the lower portion of a pile. A reduced size of coarse aggregate is often necessary for filling pile shells and should be specified accordingly. Pile shell seams should be welded unless nonwelded locked seams are proven to be capable and are approved by the authority having jurisdiction.

C4.5.5.8 Cast-in-Place Concrete Piles Concrete mix design and pile design should be considered carefully. Some types of piles require special concrete grout or grout mixes, and special mixes may be required for concrete pumped over substantial distances. Concrete aggregates, cements, polymers, admixtures, and other concrete additives, including fibers, should be appropriate for the installation procedure and intended use.

C4.5.6 Pile Design

C4.5.6.1 Pile Capacity For pile sections that are freestanding with bottom embedment, the member can be treated as a column or a vertical cantilever beam having an unbraced length:

$$L = H + d/12$$

where

L = unbraced length in feet
H = height of member in feet from top of grade, plus depth of local scour, erosion, and liquefaction, to the underside of cap or other floor structure above
d = depth of embedment, in inches, defined as the distance from top of grade minus depth of local scour and liquefaction to the point of fixity, computed conservatively as

$$d = 1.8(EI/n_h)^{1/5}$$

where

E = the modulus of elasticity of the pile (lb/in.^{2})
I = minimum moment of inertia of pile (in.^{4})
n_h = coefficient of horizontal subgrade reactions (lb/in.^{3}).

An appropriate effective length factor K, a function of end restraint, shall be applied to the unbraced length L to obtain an effective length $K(L)$.

In some situations the use of the coefficient of horizontal subgrade reaction may be inappropriate because of long-term creep effects. Other considerations for horizontal subgrade reactions should be made per a geotechnical investigation.

Minimum lateral resistance of an individual pile should be at least 5% of the design axial load because of inevitable eccentricities, lateral forces, and being out of plumb.

C4.5.6.2 Capacity of the Supporting Soils The distribution of subgrade reaction on a pile depends on both the ratio of the pressure on the pile to a corresponding displacement and the flexural rigidity of the pile. The computations required to estimate these are cumbersome and subject to error. The error is increased with smaller design lateral loads and larger assumed soil diameters (diameter of subsoils centered on the pile that can be assumed available to react against a pile). In lieu of more exact analyses, the soil diameter is limited to three times the diameter of the pile.

The capacity of supporting soils is also affected by the pile installation method. Installation methods that include jetting or augering will initially reduce the capacity of the soil surrounding the pile to the depth of the jetting or augering, and this effect will persist for some time into the future until the soil can reconsolidate around the pile. There is uncertainty on both the extent of loss of supporting soil and on the time required for reconsolidation. Pile capacity formulas can account for the reduction in capacity for these installation methods.

C4.5.6.3 Minimum Penetration Additional guidance on penetration and installation aspects of pile design can be found in ASCE 20 (1997).

C4.5.6.4 Foundation Pile Splicing Splicing of piles is difficult and, if not done properly, can compromise the structural integrity of a foundation. Given the uncertainties related to scour of erodible soils in areas subject to high velocity wave action and in High Risk Flood Hazard Areas, splicing of wood piles should be avoided. In instances where wood piles must be spliced, the splice should be below the point of fixity defined in Section C4.5.6.1. This will locate the splice in a section of pile not subject to large bending stresses. In coastal areas, pile splices should be below −10 ft MWL unless detailed analysis (including erosion and scour effects) shows a splice above this elevation will not compromise the foundation performance.

C4.5.7 Posts, Piers, and Columns Column-supported foundations can be anchored against pullout through the use of rock anchors or by penetration and grouting into nonerodible material. Where penetration into a nonerodible material is chosen, the hole into the nonerodible material should be bell-shaped and filled with grout or concrete such that the lower end of the mass will be wider than the upper, and so that pullout will be resisted even with minimal shrinkage of the grout. It is highly recommended that columns be supported on piles in this area. Field experience has shown that many columns are poorly constructed and are subject to failure.

Because of frequent difficulties in quality control of site-constructed piers, posts, and columns in residential development, wood or pre-cast concrete piles are recommended in lieu of posts, piers, and columns. Posts, piers, and columns may not be desirable in erosion-prone or wave-prone buildings and structures.

Footings that support columns usually must be extremely large to resist overturning caused by water, wind, or seismic forces. Footing sizes can approach a practically continuous footing or mat at the bases of the columns.

C4.5.7.3 Reinforced Concrete Columns Post-disaster structural assessments and field inspections have observed that small diameter columns frequently fail, either by breaking or by poor connection to footings, pile caps, and similar elements. In addition, post-disaster investigations have identified that reinforcing steel splice lengths are frequently found to be less than required by ACI 318, *Building Code Requirements for Structural Concrete and Commentary* (ACI 2011); thus, quality control inspections on reinforcing steel installation for columns should be considered.

C4.5.8 Footings, Mats, Rafts, and Concrete Slabs That Support Columns or Walls Footings may not be desirable in erosion-prone or wave-prone buildings and structures. The top of the footing must be located below the eroded ground elevation. Designers are advised that excavating to place these elements may disturb the soil enough to exacerbate local scour, and deeper embedment may be appropriate.

C4.5.10 Grade Beams In High Risk Flood Hazard Areas and in areas subject to high velocity wave action, the use of grade beams may increase the forces transmitted from the foundation to the structure and may increase scour around the foundation. However, they are often used in engineered structures to strengthen piling or other vertical members during the most severe flood conditions. The increased scour and increased loads should be mitigated by proper design of the pile system, grade beam, and structural frame. The grade beam should be designed to be self-supporting between piles or other vertical supporting elements, assuming the soil under the grade beam has been fully eroded.

C4.5.11 Bracing Cross bracing perpendicular to the primary wave and hydrodynamic forces has been limited to rods and cables to reduce the cross-sectional area of bracing presented to flood forces, and to reduce debris trapping and likelihood of bracing failure. In designing cross bracing, the designer should consider the likelihood of debris buildup against the bracing, which will increase the surface area for wave forces to act on and increase the loads to be resisted.

In general, it is recommended that the designer strive to establish a stable design that limits the reliance on bracing elements for rigidity and that minimizes use of bracing subject to flood loads. Past experience has shown cross bracing often fails during a storm event and does not provide the degree of structural support anticipated. Knee bracing has been shown to be less vulnerable to flood forces and is recommended as the first type of bracing to be considered.

When using steel rods for cross bracing in a highly corrosive or marine environment, the rods should be hot-dipped galvanized or fabricated from noncorrosive material in accordance with Chapter 5 of this standard.

C4.5.12 Shear Walls Shear walls are often required to resist lateral forces on tall buildings and structures. Where possible, these walls should be oriented such that flood load transfer to the rest of the structure, including the foundation, is minimized. Where feasible, shear walls should be built in segments such that there are breaks in the wall that will allow floodwater to flow easily around the walls. If segmentation is not feasible, the designer may evaluate whether specifying openings would relieve some flood load. For low-rise buildings and structures in V Zones, shear walls may act as obstructions that compromise structural integrity and should be avoided. See FEMA Technical Bulletin 5, *Free-of-Obstruction Requirements for Buildings Located in Coastal High Hazard Areas* (FEMA 2008c) for additional guidance.

C4.5.13 Stem Walls This standard prohibits use of stem walls in Coastal High Hazard Areas (Zone V).

This standard permits use of stem wall foundations in Coastal A Zones where soil conditions limit the use of deep foundations and where shallow foundations are permitted by Section 4.5.1.2.

Because stem wall foundations are designed to retain soil, flood openings are not required.

C4.6 ENCLOSED AREAS BELOW DESIGN FLOOD ELEVATION

The NFIP requires that the use of enclosed areas below the design flood elevation (DFE) be restricted to parking of vehicles, access, or storage; lower areas must not be finished or used for any other purpose. No mechanical, electrical, or plumbing equipment, exclusive of risers and underground lines, should be installed below the DFE.

Though the NFIP places no restrictions on the size of enclosed areas below DFE, the designer is cautioned that enclosed areas that are 300 sq ft and larger can affect flood insurance premiums, even if the walls are designed to break away in accordance with this standard. The designer may wish to contact an insurance agent to obtain information regarding enclosures and additional insurance premium charges.

An exterior door is required at the entry at the top of stairways that are enclosed by breakaway walls, if the stairway provides direct access to the elevated building. This requirement does not apply if the stairway provides access to a deck or similar feature exposed to the elements. The purpose of this requirement is to prevent the interior of the building from being subject to wave splash and wind-driven rain if the breakaway walls surrounding the stairway fail as intended. The door at the top of the stairway should meet building code requirements for testing and labeling of exterior doors. Designers should provide flashing and other details consistent with exterior door assemblies.

For enclosed areas with elevators, see Section 7.5.

C4.6.1 Breakaway Walls The breakaway feature can be accomplished either through material failure or through failure of connections holding the walls in place. Lattice and screening will break away through material failure. Where connection failure is relied on, care should be taken to ensure that individual panels will not lead to unacceptable flood and debris loads being transferred to the main structure or to other structures. In addition, wall panels should be detailed such that damage to the exterior siding of elevated buildings will not occur when the panels break away as designed. Designers are referred to the FEMA Technical Bulletin 9, *Design and Construction Guidance for Breakaway Walls Below Elevated Buildings in Coastal High Hazard Areas* (FEMA 2008a).

Fixed slats (vertical or slanted) or fixed louvers that permit floodwaters to flow through prior to material failure are not specified by the standard but may be permitted by the authority having jurisdiction. Guidance in FEMA Technical Bulletin 9, *Design and Construction Guidance for Breakaway Walls Below Elevated Buildings in Coastal High Hazard Areas* (FEMA 2008a) indicates materials should be cosmetic in nature (i.e., not structural or used for bracing), and the material used for the slats or louvers should be of small thickness. Whether lattice, fixed slats, or fixed louvers are used, the ratio of open area to the total wall covered by such materials area should be at least 40%.

C4.6.2 Openings in Breakaway Walls This standard requires flood openings in breakaway walls in V Zones and Coastal A Zones and specifies that the openings meet the same requirements as openings in A Zone enclosures. The requirement for flood openings in V Zone breakaway walls is new to this edition of the standard. Postflood experience shows that breakaway walls fail under water depths and wave conditions that may be considerably less than the water depths and waves expected during the base flood. Having openings will permit the water level inside to match the water level outside, limiting failure under "shallow" flooding that occurs more frequently than the base flood. Since breakaway walls are not covered under federal flood insurance policies, the addition of openings should save property owners money during less-than-base-flood events.

C4.7 EROSION CONTROL STRUCTURES

Erosion control structures that are attached to building foundations will increase flood loads acting on those foundations by transferring flood loads from a wide area onto the foundation. For this reason, attachment is prohibited. The presence of an erosion control structure close to a building can also increase local erosion and flood forces, so these effects should be considered in the siting and design of the erosion control structure and in the design of the building foundation and elevated building elements that could be subject to wave reflection and runup impacts. Some guidance on erosion control structure impacts and siting in close proximity to buildings is contained in FEMA Technical Bulletin 5, *Free-of-Obstruction Requirements for Buildings Located in Coastal High Hazard Areas* (FEMA 2008c).

CHAPTER C5
MATERIALS

C5.1 GENERAL

All materials used below the elevations required in Table 5-1 are to be flood damage-resistant materials. Designers should consider using such materials above the minimum elevation in order to reduce the likelihood of damage because of wetting and wicking.

Some sources of information to assist in the selection of materials for flood resistant design and construction include FEMA Technical Bulletin 2, *Flood Damage-Resistant Materials Requirements for Buildings Located in Special Flood Hazard Areas* (FEMA 2008b); Fact Sheet 1.7 of FEMA P-499, *Home Builder's Guide to Coastal Construction: Technical Fact Sheet Series* (FEMA 2010b); and *Flood Proofing Regulations* (USACE 1995).

Assemblies, such as doors, garage doors, flood opening devices, and other assemblies, should be manufactured with flood damage-resistant materials so that the assemblies can withstand contact with floodwaters without sustaining damage that requires more than cleaning and cosmetic repair. Field testing of some residential building envelope systems has been conducted by Tuskegee University and Oak Ridge National Laboratory (Aglan et al. 2004).

Designers should be aware that material testing standards are under development that will help clarify "acceptable" and "unacceptable" ratings of materials listed in FEMA Technical Bulletin 2, *Flood Damage-Resistant Materials Requirements* (FEMA 2008b) and provide a method for evaluating materials not listed in Technical Bulletin 2.

C5.2 SPECIFIC MATERIALS REQUIREMENTS FOR FLOOD HAZARD AREAS

C5.2.1 Metal Connectors and Fasteners Exposed metal connectors and fasteners should be designed to a higher safety or load factor in coastal areas susceptible to salt spray to account for loss of material because of corrosion.

Protection of metal connectors and fasteners from corrosion is accomplished either by use of corrosion resistant materials or hot-dip zinc galvanized coating. The severity of exposure to corrosive conditions is a consideration in the selection of materials and coating thickness, as discussed in FEMA Technical Bulletin 8 *Corrosion Protection for Metal Connectors in Coastal Areas* (FEMA 1996). In highly corrosive environments, such as near bodies of saltwater with waves that contribute to presence of salt-laden air, connectors and fasteners should (1) be either 304 or 316 alloy stainless steel, in accordance with ASTM A276 *Standard Specification for Stainless Steel Bars and Shapes* (ASTM 2010a), or (2) have thicker hot-dip galvanizing as provided in FEMA Technical Bulletin 8 *Corrosion Protection for Metal Connectors in Coastal Areas* (FEMA 1996).

Even metal parts protected by hot-dip zinc galvanized coatings have been observed to be weakened by effects of corrosion. In highly corrosive environments, periodic inspection and maintenance, including painting or replacement of connectors and fasteners, should be carried out as needed.

Chemicals in preservative treated wood have been shown to be corrosive to metal connectors (e.g., straps and hangers) and fasteners (e.g., nails, screws, and bolts). Minimum corrosion protection for treated wood applications is through use of hot-dip zinc galvanized coating, Type 304 or Type 316 stainless steel or other corrosion resistant materials, and the treated wood or connector manufacturer's recommendations to address corrosion. Use of moisture resistant membrane such as 15-lb roofing felt or an appropriate self-adhered membrane product to separate the metal connector from direct contact with the preservative treated wood is also considered to be an effective detail for reducing potential for corrosion.

C5.2.2 Structural Steel The standard specification for steel pipe piles is ASTM A252/252M (ASTM 2010b). Steel pipe piles come in three grades: Grade 1 (minimum 30 ksi yield strength), Grade 2 (minimum 35 ksi yield strength), or Grade 3 (minimum 45 ksi yield strength).

The standard specification for steel H piles and steel sheet piling is ASTM A572/A572M (ASTM 2013). Steel H-piles are usually Grade 50. Steel sheet piling is usually Grade 50, but Grade 55, 60, and 65 materials are also commonly produced by steel mills. Some designers may specify ASTM A36 steel (ASTM 2012a) but will generally get a pile produced to ASTM A572 that is marked with an ASTM A36 stamp. ASTM A690 (ASTM 2012b) steel should be used for sheet steel piling that will be immersed in saltwater. ASTM A690 can also be used for steel H piles to be immersed in saltwater (e.g., driven in the splash or tidal zone in a marine environment) but is rarely specified for such a use.

Rolled steel shapes other than H piles are typically produced to ASTM A992 (ASTM 2011). ASTM A36 steel may be available from some mills, but ASTM A572 and ASTM A992 are higher strength and more readily available, and thus, would provide a more economical pile. As with H piles and steel sheet piling, where ASTM A36 is specified, the actual pile supplied may be produced to ASTM A572 or ASTM A992 but marked as A36 steel.

Cast steel shoes conforming to ASTM A148 (ASTM 2008) should be used where high-strength steel is specified.

Exposed structural steel shapes, such as W sections, tees, plates, bars, pipes, channels, and angles used for beams, columns, bracing, and lintels, should be avoided in flood hazard areas wherever possible, especially in coastal areas. Even when protected by a galvanized coating or other means, corrosion has been observed to occur rapidly.

C5.2.3 Concrete Designers should consider specifying a minimum compressive strength of 5,000 psi for structures that are located within 3,000 ft of the coastline to increase resistance to saltwater intrusion, consistent with ACI 318 requirements, and to increase concrete resistance to salt spray and saltwater. Designers also may wish to refer to *Code Requirements for Environmental Engineering Concrete Structures*, ACI 350, *and Commentary*, ACI 350R (ACI 2006) when designing concrete structures in flood hazard areas. Although the requirements may be conservative for many applications, they may be useful for certain structures that might require or benefit from increased protection.

C5.2.4 Masonry Designers should consider specifying that all masonry units below the required lowest floor elevations be fully grouted to minimize trapping of floodwaters and moisture that can foster mold growth in spaces that cannot be cleaned after floodwaters recede.

C5.2.5 Wood The requirement for preservative treated wood in pile, post, and brace applications is necessary to protect the wood from deterioration because of decay hazards, such as ground contact and weather.

Previous editions of the standard required wood and timber members to be naturally decay resistant or pressure treated with preservatives. Starting with this edition, preservative treatment is only required where specified. Wood products that are acceptable according to FEMA Technical Bulletin 2, *Flood Damage-Resistant Materials Requirements* (FEMA 2008b), such as certain plywood panels and wood studs, have been shown to be resistant to the effects of flood exposure without the aid of preservatives or the use of naturally durable wood. Specifically, exterior structural plywood that is classified as Exposure 1 grade in accordance with APA Standard PS1 (APA 1995) has been shown to resist flood exposure effects. Structural plywood bonded with interior glue is not considered to be flood damage resistant.

Pile and post applications, which place the wood member in contact with the ground, commonly employ preservative treated wood. This material selection is largely because of the severity of the decay hazard, as well as availability of preservative treated wood species in structural sizes for pile and post applications. Whereas heartwood of naturally decay resistant species identified as having exceptionally high decay resistance per the *Wood Handbook—Wood as an Engineering Material* (Forest Products Laboratory 2010) can be used as an alternative to preservative treated wood, special considerations, such as experience from prior use and quality procedures to ensure presence of heartwood, should be made for pile and post applications where the wood is exposed to both weather and ground contact conditions. Wood posts and piles must be graded as structural lumber and not have defects that would reduce the structural capacity of the member. The lumber quality of wood posts and piles must be such that wood preservative treatment can be fully effective throughout the column area and length.

Plywood and structural wood treated with borate preservative are acceptable for flood damage resistance but are not recommended for use below the DFE for purposes of termite or decay resistance because of potential for leaching when exposed to water (see FEMA Technical Bulletin 2, FEMA 2008b).

CHAPTER C6
DRY FLOODPROOFING AND WET FLOODPROOFING

C6.1 SCOPE

Planning and design of floodproofing measures should consider the implications, including aesthetics, construction costs, alternative designs, and flood insurance premium charges of floodproofing to an elevation higher than the minimum required elevation. National Flood Insurance Program (NFIP) flood insurance for structures that are dry floodproofed to at least the BFE plus 1 ft may be rated as being protected to the BFE. Protection to higher levels may qualify buildings for lower premiums.

There are a number of documents that provide detailed information on planning, design, and construction of floodproofing measures. Among the more notable are the U.S. Army Corps of Engineers publication, *Flood Proofing Regulations* (USACE 1995) and two publications, FEMA P-936, *Floodproofing Non-Residential Buildings* (FEMA 2013a) and FEMA P-259, *Engineering Principles and Practices for Retrofitting Flood Prone Residential Buildings* (FEMA 2012a). These documents provide information for engineers, architects, local officials, and building owners regarding all aspects of floodproofing existing structures in flood hazard areas. For suggestions for existing, nonconforming residential structures, designers are referred to FEMA P-312, *Homeowner's Guide to Retrofitting: Six Ways to Protect Your House from Flooding* (FEMA 2009a).

NFIP regulations regarding floodproofing, contained in 44 CFR 60.3(c)(3), (4), and (8), should also be consulted. These regulations are discussed in FEMA Technical Bulletin 3, *Non-Residential Floodproofing—Requirements and Certification for Buildings Located in Special Flood Hazard Areas* (FEMA 1993b).

Design of levees and floodwalls intended to provide flood protection to an individual or group of structures is outside the scope of this standard. Such flood control measures do not "floodproof" buildings and structures. There are numerous engineering, non-engineering, and regulatory issues involved in the design and construction of levees and floodwalls. Therefore, caution should be exercised when considering employing a levee or floodwall as a method of providing flood protection. Local floodplain management officials and various state and federal agencies that may have regulatory authority over levees and floodwalls should be consulted.

Dry floodproofing is not permitted in Coastal High Hazard Areas (V Zone), Coastal A Zones, and in other High Risk Flood Hazard Areas (defined in the standard). Certain buildings and structures in flood hazard areas where dry floodproofing is permitted may be designed and constructed with dry floodproofing so that, together with attendant utility and sanitary facilities, areas below the minimum specified elevation are watertight with walls substantially impermeable to the passage of water and with structural components having the capability of resisting hydrostatic and hydrodynamic loads and effects of buoyancy. For dry floodproofing, flood loads should be calculated to the elevation specified in Table 6-1 of the standard so that the required level of protection is achieved.

For consistency with the NFIP, dry floodproofing is applicable only for nonresidential buildings and nonresidential portions of mixed-use buildings. This standard defines the terms "nonresidential" and "residential" in Section 1.2. The term "mixed use" is defined in Section C1.2.

For new structures, floodproofing measures built into a structure and requiring no human intervention during the flood event are the most reliable and preferred method of floodproofing. Floodproofing measures requiring human intervention inherently have a greater chance of failure. For nonconforming, existing structures, relocation to less flood-prone sites and elevation measures are most likely to successfully protect flood-prone structures; however, other floodproofing measures, when properly designed for existing structures, can also provide flood protection.

C6.2 DRY FLOODPROOFING

Whenever dry floodproofing is proposed for any area below the required elevation specified in Table 6-1, whether the area is above grade, below grade, or a combination of the two, assurance must be provided that reliable flood protection will be achieved and that the structure will be substantially impermeable to the passage of floodwater against floods up to the specified elevations. This requires strict adherence to materials and construction requirements for dry floodproofing. Designers are encouraged to specify testing of dry floodproofing measures as part of construction to identify potential problems or leaks under controlled conditions. See FEMA Technical Bulletin 6, *Below-Grade Parking Requirements for Buildings Located in Special Flood Hazard Areas* (FEMA 1993a) and *American National Standard for Flood Abatement Equipment* (ANSI/FM Approvals 2014). Sump pumps should be provided to handle inevitable seepage, and emergency power should be provided to run the pumps, especially in areas where inundation duration is expected to last more than 12 h.

Table 6-1 specifies the minimum elevation of dry floodproofing as a function of the Flood Design Class assigned to each building and structure (see Section 1.4.3 and Table 1-1 of the standard). Flood Design Classes 1, 2, and 3 require dry floodproofing to BFE plus 1 ft or DFE, whichever is higher. Flood Design Class 4, which includes essential facilities (sometimes referred to as "critical facilities"), requires dry floodproofing to BFE plus 2 ft, DFE, or the 500-year flood elevation, whichever is higher.

Designers and owners are cautioned that unlike elevated buildings that would experience relatively shallow flooding if floodwaters rise above the lowest floor, under the same flood scenario, dry floodproofing measures may be overtopped, resulting in spaces below the water level being subject to deep flooding and possible structural damage. The more important a building and its functions are, the more attention should be given to the consequences of flooding that exceeds the design conditions, especially if critical facilities cannot be located outside of flood hazard areas or in less hazard-prone areas.

Designers are advised that the authority having jurisdiction will require the designer to provide signed and sealed documentation that the design and proposed methods of construction meet the requirements of this standard for dry floodproofing. To obtain NFIP flood insurance for dry floodproofed buildings and structures, owners will be required to provide the NFIP *Floodproofing Certificate* (FEMA 2012b).

Although emergency plans are required by this standard only if removable shields are used (see Section 6.2.3), and such plans are required to specify maintenance, designers and owners are advised to develop annual inspection plans of the structural and nonstructural components of the dry floodproofing measures. Lack of maintenance of such components could contribute to inadequate performance and damage. Some authorities having jurisdiction may require periodic inspection reports or on-site inspections.

The standard distinguishes buildings and portions of buildings as either residential, nonresidential, or mixed use. The term "mixed use" is defined in Section C1.2.

Existing nonresidential buildings and nonresidential portions of mixed-use buildings may be dry floodproofed using retrofit methods. Retrofitting such an existing building with dry floodproofing measures may be considered if the building is required to be brought into compliance because of substantial improvement or repair of substantial damage, in which case the retrofit dry floodproofing measures should achieve the same performance expected of new structures that are dry floodproofed. Even if compliance is not required because of substantial improvement or substantial damage, owners and designers may use dry floodproofing retrofit measures or other retrofit measures to reduce vulnerability to future flooding, in which case this standard may be used as guidance.

Existing residential buildings and residential portions of mixed-use buildings are permitted to be retrofitted with dry floodproofing measures only if compliance is not required (i.e., not substantial improvement or substantial damage). Any residential building that was designed and built in compliance with floodplain management requirements at the time it was constructed (post-FIRM) may not be retrofitted with dry floodproofing measures if such measures would make the building noncompliant with the original requirements.

Extensive evaluation of an existing structure's ability to resist flood loads and all possible avenues of entry of water must be undertaken to ensure that retrofit dry floodproofing methods appropriately account for flood loads and failure points. Where proposed work on existing buildings, including the dry floodproofing work and all other work, constitutes substantial improvement, compliance with the requirements for new structures is required.

When building services, such as power, steam, water, sewer, fire safety, and telecommunications, are underground the designer is cautioned that achieving the desired performance requires consideration of measures to prevent or minimize water intrusion through the openings that are necessary for access by these building services. When considering retrofitting dry floodproofing of existing buildings, if it is unlikely that these building services can be rerouted above the required elevation, then flood protection is accomplished by sealing penetrations through the walls and floors. If adequate sealing cannot be accomplished, equipment may be surrounded by dry floodproofed enclosures, although the designer is cautioned that by itself this measure may not meet building code requirements when compliance is required.

C6.2.1 Dry Floodproofing Limitations Dry floodproofing of residential structures and residential portions of mixed-use structures is not permitted because of the danger to occupants should the dry floodproofing measures fail or be exceeded by flooding. In addition, dry floodproofing measures frequently require human action, such as installing flood shields, maintaining the protective features, having an operational plan, and being able to take action within a reasonable warning time. The possible inability or failure of owners or other occupants to take such an action because of absence, lack of maintenance, or change in ownership of the home is regarded as an unacceptable risk.

Because of the nature of flood loads and the possibility of increased erosion and scour along solid walls, dry floodproofing measures are not allowed for structures in Coastal High Hazard Areas and Coastal A Zones.

The 5 ft/s velocity restriction for floodproofing is not a requirement of the NFIP but is used in the U.S. Army Corps of Engineers' publication *Flood Proofing Regulations* (USACE 1995). Although effective dry floodproofing can be designed for higher velocities, this is a reasonable existing limit that addresses safety of dry floodproofed structures during a flood. Whether certified dry floodproofing designs that demonstrate resistance to higher velocities while meeting the other limitations of this standard will be accepted is a decision for the authority having jurisdiction.

C6.2.2 Dry Floodproofing Requirements For safety, dry floodproofed buildings should not remain occupied during conditions of flooding. The requirement is to provide a means of access by emergency personnel and a means of escape for occupants who do not evacuate prior to the onset of flooding. In some cases, there may be a need for a limited number of personnel to remain in a facility to secure equipment or monitor ongoing operations or conditions during the flooding. This may be the case for mid- or high-rise buildings and some industrial, healthcare, or other critical facilities, such as power generating stations and emergency operations centers.

C6.2.3 Limits on Human Intervention In some instances, dry floodproofing measures require human intervention to be effective. However, the design of measures that require human intervention should minimize the effort needed and the complexity of deployment. Additionally, adequate warning from a credible source must be available, and strict guidelines for implementing dry floodproofing measures that require human intervention must be followed (persons responsible for installing or implementing the measures must be familiar with the procedures and equipment; sufficient warning time must be given to ensure the measures are put into place, etc.). Guidance on flood warnings is available in a document entitled *Flood Warning Systems Manual* (NWS/HSP 2012).

C6.3 WET FLOODPROOFING

Wet floodproofing techniques are used to reduce flood damage when an enclosed area of a structure is designed to allow entry and exit of floodwaters. However, all construction materials below the DFE should be flood damage-resistant materials (see Chapters 5 and C5) and have appropriate structural strength to resist flood forces. FEMA Technical Bulletin 7, *Wet Floodproofing Requirements for Structures Located in Special Flood Hazard Areas* (FEMA 1993c), provides information on NFIP regulations regarding wet floodproofing.

The areas for which wet floodproofing is permitted include Flood Design Class I structures; enclosed areas that are below elevated structures and that are used solely for building access, parking of vehicles, and storage; functionally dependent structures, defined in the standard; and certain types of agricultural structures. Some examples of enclosed areas that may be wet floodproofed include garages, crawlspaces, stairwells, and storage areas.

CHAPTER C7
ATTENDANT UTILITIES AND EQUIPMENT

C7.1 GENERAL

To avoid damage by floodwaters, electrical, plumbing, mechanical, HVAC, and other systems that serve buildings and structures are to be located or elevated above the minimum elevations in Table 7-1. Platforms used to elevate equipment are to comply with the standard. When installed on platforms that are outside of buildings, equipment should also be anchored to resist wind, ice, snow, rain, and seismic loads.

Alternatively, if located below the minimum elevations in Table 7-1, attendant utilities and equipment are to be specifically designed and protected to resist damage by preventing floodwater from entering or accumulating within them. Another alternative available for nonresidential buildings and structures is to locate the systems and components within enclosed structures or portions of structures that are dry floodproofed in accordance with Section 6.2.

Flood damage to attendant utilities and equipment may result in effluent discharge from sewer lines, release of fuel from tanks, contamination of potable water, saltwater intrusion into attendant utilities and equipment, fire hazards from damage to gas or electric appliances and lines, and other preventable losses. Further, damage to equipment and appliances or disruption of utility services can delay reoccupancy after a flood, even if structures are otherwise undamaged.

The provisions of this chapter apply to all attendant utilities and equipment unless the authority having jurisdiction specifically determines that some elements necessary for fire and life safety purposes must be located below the elevations in Table 7-1. This chapter is intended to apply to installation of emergency power systems, including generators.

FEMA 348, *Protecting Building Utilities from Flood Damage: Principles and Practices for the Design and Construction of Flood Resistant Building Utility Systems* (FEMA 1999) outlines guidance from the National Flood Insurance Program (NFIP) and methods to achieve the performance necessary to minimize flood damage.

In addition to the stated requirements, portions of utility systems that extend below the required elevation, such as piping, are to be designed to resist anticipated flood loads and erosion and scour expected during the design flood conditions. If portions of these systems are in protective floodproofed enclosures that are attached to structures, the enclosures and the structures are to be designed for flood loads, including loads transmitted to the structures. FEMA P-55, *Coastal Construction Manual* (FEMA 2011) provides more specific guidance on designing in the coastal environment, including the design of supporting exterior platforms for utilities and attendant mechanical equipment.

Where components of attendant utilities and equipment are allowed below the elevation in Table 7-1 and where exposed to hydrodynamic loads, flood-borne debris, or wave loads, a good practice is to place such installations on the most sheltered side of foundation members. Analysis of potential flood sources and pathways will help to identify the most sheltered side.

In all cases, utilities and conduits, cables, switches, pipes, faucets, and other such attendant components are not to be located on or penetrate through walls designed to break away. Postflood investigations indicate that such installations adversely affect the ability of walls to break away as designed.

Recent postflood investigations have identified vulnerabilities due to attendant utilities and equipment, including emergency generators and their fuel tanks and associated equipment needed for backup electric service that were located either outdoors at ground level or in areas of buildings and structures that were subject to flooding. Although observed primarily in older buildings, especially mid- and high-rise buildings, vulnerable installations were also observed in newer buildings that should have conformed to utility and equipment requirements for flood hazard areas. Even if compliance with this chapter is not required, designers working with owners of existing buildings are encouraged to evaluate whether utilities and equipment can be relocated, elevated, or otherwise protected from flooding in conformance with this chapter.

This chapter is not intended to apply to underground or aboveground electric power lines that bring power to buildings, water supply distribution lines that bring treated water to buildings, or to sanitary sewage systems that take effluent from buildings. States and communities may have requirements for water and sewer systems in flood hazard areas.

C7.2 ELECTRICAL SERVICE

Installations of certain components of electrical service that are exterior to structures may be regulated by state public service commissions and may not be subject to local requirements. Whether electrical service is distributed underground or overhead, flood damage is minimized and service can be more rapidly restored if meter bases, panelboards, load centers, disconnect switches, and circuit breakers are elevated above the elevation required in Table 7-1. If elevating these components makes access difficult, a low platform or an automated meter reading system may be appropriate. Similarly, damage is minimized if transformers, switchgear, and other exterior equipment associated with electrical distribution are located above the design flood elevation (DFE).

For underground distribution, transformers are frequently pad-mounted and located below the DFE. Designers and building owners should consider requesting that local utility companies elevate transformers that serve buildings in flood hazard areas. The National Electrical Code (Standard Reference NFPA 2011) should be consulted regarding burial requirements for electrical lines.

C7.2.5 Electric Elements Installed below Minimum Elevations If required to meet the life safety provisions of the code or if the authority having jurisdiction deems electric elements, such as circuits, switches, receptacles, and fixtures, to be expendable, thus may be allowed below the elevation required in Table 7-1, then the number of such electric elements should be kept to a minimum, and such electric elements should be intended for wet locations. Except for some low-voltage applications, conductors are not considered expendable. In addition, branch circuits originating from ground-fault circuit-interrupter breakers should be used to supply any electric element allowed below the elevation required in Table 7-1.

C7.3 PLUMBING SYSTEMS

C7.3.3 Plumbing Systems Installed below Minimum Elevations Certain system components (such as piping) and fixtures (such as outdoor faucets and showers) may be located below the DFE and meet the requirement to prevent the release of sewage into floodwaters and prevention of infiltration of floodwaters into the plumbing system (especially potable water). However, other components and fixtures (such as toilets and sinks) are not allowed. Backwater or backflow prevention devices may be used, for example, for floor drains but are not to be used to allow plumbing installations below the required elevation unless otherwise allowed by this section. In addition, plumbing system components and fixtures should not be installed in enclosures below elevated buildings, because the uses of such enclosures are to be restricted to building access, parking, and limited storage.

C7.3.4 Sanitary Systems Designers are advised to consult with the appropriate regulatory authority, an experienced sanitation expert, or other professional regarding design and installation of on-site sanitary systems in order to provide adequate functioning without adverse health effects during flood conditions.

Post-disaster inspections have frequently identified failures in sanitary systems, including septic tanks dislodged by buoyant forces, underground tanks filled with floodwater and debris, interior drains that have backed up into buildings from overflowing sewage disposal systems, and drain system collapse from saturated soil covering the drain field.

C7.4 MECHANICAL, HEATING, VENTILATION, AND AIR CONDITIONING SYSTEMS

Rather than place ductwork below the elevations in Table 7-1, designers are advised to consider additional elevation of the floor system so that underfloor ductwork is above the elevations in Table 7-1. However, if ductwork and duct insulation are placed below the specified elevation, the ductwork is to be designed, constructed, and installed to resist flood loads, as well as to prevent floodwaters from entering into and accumulating within the ductwork and is to be constructed of flood resistant materials. Duct insulation generally is not resistant to flood damage. In addition, when saturated, insulation exerts considerable load on hanger straps and connections that tend to fail, resulting in significant damage to the ductwork.

C7.5 ELEVATORS

The intent of this section is to prevent flood damage to any significant elevator component, where "significant" can be considered to include any structural components, any expensive components, or any components that are difficult (or expensive) to access or replace or restore. Thus, any significant elevator components shall be either (1) elevated above the elevation specified in Table 7-1, or (2) constructed of flood damage-resistant materials and be capable of resisting physical damage due to design flood conditions. The phrase "physical damage due to design flood conditions" is used in this section to indicate damage from any source during the design flood—hydrostatic, hydrodynamic, waves, flood-borne debris, sedimentation or erosion, corrosion, and so on. Components below the elevations specified in Table 7-1 are not required to be flood damage resistant if (1) they are deemed not significant by the authority having jurisdiction, and (2) they are easy to access and inexpensive to replace or clean.

FEMA Technical Bulletin 4, *Elevator Installation for Buildings Located in Special Flood Hazard Areas* (FEMA 2010a) provides guidance related to elevator installations in flood hazard areas. Much of the following text was taken from that document.

There are two primary types of elevators: hydraulic and electric traction. The hydraulic elevator consists of a car attached to the top of a hydraulic jack assembly, which normally extends below the lowest floor and is operated by hydraulic pressure. The electric traction elevator, the more common type found in structures over three or four stories, is raised by suspension cables attached to a car and counterweights that pass over a traction sheave driven by an electric motor generally located above the elevator hoistway.

Hydraulic elevators typically require part of the assembly to be located below the DFE because the jack assembly is located below all floors that are served by the elevator. The hydraulic pump and reservoirs, however, can easily be located up to two floors above the jack, thus usually above the elevations specified in Table 7-1. Designers are advised that avoidance of inundation is important in coastal communities because saltwater is particularly damaging because of its corrosive nature.

Potential flood damage to elevators can be dramatically reduced when using traction elevators, as the majority of the equipment is normally located above the elevator shaft and, therefore, not exposed to floodwaters. When some equipment cannot be located above the elevations specified in Table 7-1, such as oil buffers, compensation cable, and governor rope sheaves, which usually are located at the bottom of the shaft, flood resistant materials should be used where possible.

To minimize loss of life, injuries, and damage, and to ensure elevators are serviceable after flooding, a system of interlocking controls with one or more float switches in the elevator shaft to prevent the cab from descending into floodwaters should be installed. Owners should consider permanent signs in elevators and at call buttons advising users of elevator operations in the event of flooding. In addition, the majority of the electrical equipment for both elevator types, such as electrical junction boxes and circuit and control panels, should be elevated above the elevations specified in Table 7-1.

C7.5.1 Elevator Shafts Although use of breakaway walls is not required in areas subject to wave action, rigid, nonbreakaway walls that form elevator shafts can transmit significant loads to building foundations. Loads may be reduced by use of breakaway walls.

Designers and owners should be aware that NFIP flood insurance policies provide coverage for elevators and related equipment, although coverage is limited for equipment located below elevated buildings. The presence of elevators and shafts, and the manner in which they are constructed, are factors used to determine an insurance "rate loading," which increases the cost of premiums. Flood insurance premiums will be higher in A Zones if elevator shafts do not have flood openings and in V Zones if nonbreakaway walls are used.

CHAPTER C8
BUILDING ACCESS

C8.1 GENERAL

Buildings and structures in flood hazard areas, even if designed to resist flood loads, are not intended to be occupied during flooding events. However, some critical and essential facilities are likely intended to remain functional and occupied. In addition, it is unrealistic to expect multistory residential buildings in urban areas to fully evacuate before the onset of flooding, given the limited number of alternative shelters and likely limited private transportation for evacuation. See Section C6.2.2 about access to dry floodproofed buildings that must (or might) be occupied during flooding. Any other buildings and structures that are likely to be occupied during a flood event should be designed to provide a means of access by emergency personnel and a means of escape for occupants.

Exterior stairs and ramps located in Coastal High Hazard Areas and Coastal A Zones should be designed to transfer minimal loads to the foundation and structure supported above. Orientation parallel to the direction of flow, and open risers and railings that permit water to flow through or are designed to break away, and stairs and ramps that can be raised or retracted, are just a few methods to reduce or eliminate loads that would otherwise be transferred to the rest of the structure.

When intended to break away, stairs and ramps, and the walls that enclose them, if any, should be designed and constructed to fail without damaging the structure to which they are attached. Breakaway components should be designed to break into small pieces to avoid becoming debris that has the potential to cause structural damage to other structures.

An alternative to traditional in-place stair and ramp design is to use retractable stairs and ramps that can be retracted or raised above the elevations specified in Tables 2-1 and 4-1 prior to the onset of flooding. Because some flood events may be concurrent with high wind events, retractable ramps or stairs that are raised should resist wind loads in the retracted or raised position. All stairs and ramps should be designed in compliance with local building codes.

Foyers and enclosed or unenclosed stairways and ramps may be located under elevated buildings to provide access. If an enclosed area is designed and built to meet the applicable requirements of this standard, then the enclosure is not the lowest floor for the purpose of flood insurance rating.

Section 4.6 of this standard requires exterior doors to be used at the building entry at the top of stairs that are enclosed with breakaway walls to provide additional protection to the interior of the building from wave splash and high winds and wind-driven rain in the event such enclosures break away as intended.

CHAPTER C9
MISCELLANEOUS CONSTRUCTION

Many of the requirements in this chapter were developed on the basis of observed performance of miscellaneous construction associated with low-rise residential structures in coastal areas. The requirements can be applied to larger buildings and structures but may not include the variety of miscellaneous construction associated with larger buildings and structures, residential and nonresidential. In such cases, the general performance requirements listed in this chapter and in other chapters should guide design and construction.

Guidance for miscellaneous construction such as decks, pools, and accessory structures is contained in publications such as FEMA P-499, *Home Builder's Guide to Coastal Construction: Technical Fact Sheet Series* (FEMA 2010b) and FEMA P-55, *Coastal Construction Manual* (FEMA 2011). Although written for coastal environments, these publications can be applied to miscellaneous construction in riverine flood hazard areas.

C9.2 DECKS AND PORCHES

In flood hazard areas subject to waves and fast-moving waters, results of post-storm evaluations and surveys have indicated that decks and porches are often founded on insufficiently embedded posts, columns, or piles. The decks or porches can be undermined, transferring loads to the primary structures, or they can collapse and become flood-borne debris, which can increase debris damage to structures.

Decks and porches enclosed by lattice or insect screening are not required to meet the requirements for enclosed areas (see Sections 2.7 and 4.6). If enclosed with solid, rigid walls, they should meet the use limitations and requirements of Section 9.4.

C9.2.1 Attached Decks and Porches Attached decks and porches should be supported by foundations meeting the same requirements as the main structure or designed and constructed as cantilevers.

C9.2.2 Detached Decks and Porches Where decks and porches are detached, they should be designed and situated so as not to adversely affect the primary structure and adjacent structures through the diversion of floodwaters.

C9.3 CONCRETE SLABS

Types of concrete slabs listed in the standard should be designed and constructed so as not to cause damage to adjacent buildings and structures. Outside of Coastal High Hazard Areas, Coastal A Zones, and High Risk Flood Hazard Areas, this standard imposes no specific requirements for concrete slabs.

In Coastal High Hazard Areas, Coastal A Zones, and High Risk Flood Hazard areas, concrete slabs must meet one of two specific requirements, the slabs should either (1) be designed as frangible slabs, not structurally connected to the main structure and not capable of producing concrete debris that will cause significant damage to other structures, or (2) remain in place and functional after the design flood.

The first requirement is in response to post-disaster investigations in coastal areas that conclude reinforced slabs and thick slabs can damage building foundations if the slabs are displaced. Similar problems have not been observed with thin, unreinforced concrete slabs designed to break into small pieces if undermined.

The second requirement calls for self-supporting slabs that can meet their intended purpose whether or not they are undermined by erosion and scour. These slabs are designed to function with or without underlying soil and are often integral to the building foundation. In such cases, the building foundation must be designed to support lateral and vertical loads imposed by the slab during design flood conditions, including the effects of erosion and scour.

Stone or masonry pavers used to finish plazas and similar outdoor surfaces should be bedded in a substrate. Loose-laid pavers may be dislodged during flood conditions and damage adjacent structures, including glass doors and wall panels.

C9.4 GARAGES, CARPORTS, AND ACCESSORY STORAGE STRUCTURES

This standard defines "accessory storage structure" but places no size or cost limitations on such structures or garages. Designers should consult with authorities having jurisdiction to determine if size, cost, or other limits have been established. FEMA guidance for accessory structures associated with dwellings suggests that these accessory structures should be "small" (e.g., 100 sq ft or smaller) and "low cost" (e.g., $1,000 or less). See FEMA Technical Bulletin 7, *Wet Floodproofing Requirements* (FEMA 1993c) and FEMA Technical Bulletin 5, *Free-of-Obstruction Requirements* (FEMA 2008c).

C9.4.1 Attached Garages, Carports, and Accessory Storage Structures Attached garages, attached carports, and attached accessory storage structures may be located adjacent to or under elevated buildings, provided the primary structures and the attached structures are designed for the anticipated loads. If the attached structures are designed and built to meet the applicable requirements for enclosures below the DFE, then the floor of the garage, carport, or accessory storage structure is not the lowest floor for the purpose of flood insurance rating.

For flood hazard areas other than Coastal High Hazard Areas, Coastal A Zones, and other High Risk Flood Hazard Areas, the requirements include flood openings, flood resistant materials below the DFE, and elevated utilities and equipment. Adequate entry and exit of floodwaters are not ensured by gaps between the door segments and the garage door and the garage door jamb.

In addition, opening requirements are not satisfied when human judgment and action are needed to open the garage door or other doors prior to flooding.

Below-grade areas, including garages, are not permitted in residential structures. For nonresidential structures and the nonresidential portions of mixed-use structures, garages may be designed and constructed with below-grade areas provided the structures are not located in Coastal High Hazard Areas, Coastal A Zones, and other High Risk Flood Hazard Areas, and provided they are floodproofed in accordance with this standard, FEMA Technical Bulletin 6, *Below-Grade Parking Requirements* (FEMA 1993a) and FEMA Technical Bulletin 3, *Non-Residential Floodproofing—Requirements and Certification for Buildings Located in Special Flood Hazard Areas* (FEMA 1993b).

One of the critical elements in the dry floodproofing design of a below-grade garage is the point where the garage entrance ramp meets the street grade. The optimum design would be to design the entrance at or above the elevation required in Table 6-1. However, since the garage entry must often meet street grade elevations that are below that elevation, if the limitations on human intervention outlined in Section 6.2.3 are satisfied, the garage can be designed with a high-strength flood shield that can withstand the anticipated hydrostatic pressures and duration of flooding to prevent floodwaters from entering the dry floodproofed garage. In addition, a sufficient number of emergency exits should be available so that anyone in the garage will not be trapped by rising floodwaters.

C9.5 CHIMNEYS AND FIREPLACES

Chimneys and fireplaces are typically either site-built masonry or factory-built. If not designed to account for flood loads, and if not supported on adequate foundations, chimneys and fireplaces may be undermined by fast-moving waters or waves that may result in erosion and scour, thus causing the chimney or fireplace to crack or separate, which can damage the primary structure. Designers should consider elevating chimneys and fireplaces to match the lowest floor elevation.

C9.6 POOLS

Saturated soils and buoyant forces may dislodge, float, or damage pools unless the pools are full of water. In areas where pools generally are empty part of the year, designers should determine flood loads assuming pools will be empty.

Designers and owners are advised to review the limitations and requirements for enclosed areas below the design flood elevation (see Sections 2.7 and 4.6). The use of such enclosures is limited to only parking of vehicles, storage, and building access. Pools and other recreational uses do not meet the limitations on use.

In flood hazard areas other than Coastal High Hazard Areas, Coastal A Zones, and High Risk Flood Hazard Areas, (1) pools not surrounded by walls are permitted under elevated buildings, and (2) pools surrounded by walls are permitted in nonresidential buildings (and nonresidential portions of mixed-use buildings) and below the design flood elevation, provided the portion of the building in which the pool is located, if below the design flood elevation, meets the dry floodproofing requirements of Section 6.2.

In Coastal High Hazard Areas and Coastal A Zones, pools under elevated buildings are allowed but are not permitted to be surrounded by walls, even walls designed to break away under flood loads.

C9.7 TANKS

This section applies to all tanks, including those associated with utility service to a building or that are attached to or located under a building, and tanks that do not serve buildings, such as chemical and fuel tanks. Previous editions of this standard included provisions for tanks that support utilities for buildings in Chapter 7 and other tanks in this chapter.

Post-disaster investigations have determined that flood damage sustained by pumps, vents, and other equipment associated with fuel tanks contributes to emergency power generator failure if those elements are not located above the DFE or otherwise protected.

The factor of safety of 1.5 is a traditional value in flood resistant design, and may differ from that in other design standards that apply to tanks. Conflicts or differences between this standard and other applicable standards should be resolved with the authority having jurisdiction such that the performance requirements of the NFIP are equaled or exceeded, including consideration of the type and size of tank for which the factor of safety is intended.

CHAPTER C10
REFERENCES

All applicable documents referenced in the commentary are listed below with full titles and dates.

American Concrete Institute (ACI). (2004). "Guide for concrete floor and slab construction." *ACI 302.1R*, Farmington Hills, MI.

ACI. (2006). "Code requirements for environmental engineering concrete structures and commentary." *ACI 350*, Farmington Hills, MI.

ACI. (2011). "Building code requirements for structural concrete and commentary." *ACI 318*, Farmington Hills, MI.

Aglan, H., Wendt, R., and Livengood, S. (2004). "Field testing of energy-efficient flood-damage-resistant residential envelope systems summary report." *ORNL/TM-2005/34*, Tuskegee University, Tuskegee, AL, and Oak Ridge National Laboratory, Oak Ridge, TN.

American National Standards Institute and FM Approvals (ANSI/FM Approvals). (2014). "American national standard for flood abatement equipment." *ANSI/FM Approvals 2510*, Washington, DC <https://www.fmglobal.com/assets/pdf/fmapprovals/2510ansi.pdf> (October 6, 2014).

American Society of Civil Engineers (ASCE) (1997). "Standard guidelines for the design and installation of pile foundations." *ASCE 20-96*, New York.

American Society for Testing and Materials (ASTM). (2008). "Standard specification for steel castings, high strength, for structural purposes." *A148/A148M*, ASTM International, West Conshohocken, PA.

ASTM. (2010a). "Standard specification for stainless steel bars and shapes." *A276*, ASTM International, West Conshohocken, PA.

ASTM. (2010b). "Standard specification for welded and seamless steel pipe piles." *A252/A252M*, ASTM International, West Conshohocken, PA.

ASTM. (2011). "Standard specification for structural steel shapes." *A992/A992M*, ASTM International, West Conshohocken, PA.

ASTM. (2012a). "Standard specification for carbon structural steel." *A36/A36M*, ASTM International, West Conshohocken, PA.

ASTM. (2012b). "Standard specification for high-strength low-alloy nickel, copper, phosphorus steel H-piles and sheet piling with atmospheric corrosion resistance for use in marine environments." *A690/A690M*, ASTM International, West Conshohocken, PA.

ASTM. (2013). "Standard specification for high-strength low-alloy columbium-vanadium structural steel." *A572/A572M*, ASTM International, West Conshohocken, PA.

Bowles, J. E. (2001). *Foundation analysis and design*. McGraw Hill: New York.

Engineered Wood Association (APA). (1995). "Voluntary product standard PS1-95, construction and industrial plywood (with typical APA trademark)." Tacoma, WA.

Federal Emergency Management Agency (FEMA). (1987). "Reducing losses in high risk flood hazard areas: A guidebook for local officials." *FEMA 116*, FEMA, Washington, DC <http://www.fema.gov/library/viewRecord.do?id=1508> (October 6, 2014).

FEMA. (1989). "Alluvial fans: Hazards and management." *FEMA 165*, FEMA, Washington, DC <http://www.fema.gov/library/viewRecord.do?id=1633> (October 6, 2014).

FEMA. (1993a). "Below-grade parking requirements for buildings located in special flood hazard areas." *Technical Bulletin 6*, FEMA, Washington, DC <http://www.fema.gov/media-library/assets/documents/3498?id=1719> (October 6, 2014).

FEMA. (1993b). "Non-residential floodproofing—Requirements and certification for buildings located in special flood hazard areas." *Technical Bulletin 3*, FEMA, Washington, DC <http://www.fema.gov/media-library/assets/documents/3473?id=1716> (October 6, 2014).

FEMA. (1993c). "Wet floodproofing requirements for structures located in special flood hazard areas." *Technical Bulletin 7*, FEMA, Washington, DC <http://www.fema.gov/media-library/assets/documents/3503?id=1720> (October 6, 2014).

FEMA. (1996). "Corrosion protection for metal connectors in coastal areas." *Technical Bulletin 8*, FEMA, Washington, DC <http://www.fema.gov/media-library/assets/documents/3509?id=1721> (October 6, 2014).

FEMA. (1999). "Protecting building utilities from flood damage: Principles and practices for the design and construction of flood resistant building utility systems." *FEMA 348*, FEMA, Washington, DC <http://www.fema.gov/library/viewRecord.do?id=1750> (October 6, 2014).

FEMA. (2000). "Above the flood: Elevating your floodprone house." *FEMA P-347*, FEMA, Washington, DC <http://www.fema.gov/library/viewRecord.do?id=1424> (October 6, 2014).

FEMA. (2001a). "Crawlspace construction for buildings in special flood hazard areas." *Technical Bulletin 11*, FEMA, Washington, DC <http://www.fema.gov/media-library/assets/documents/3527?id=1724> (October 6, 2014).

FEMA. (2001b). "Ensuring that structures built on fill in or near special flood hazard areas are reasonably safe from flooding." *Technical Bulletin 10*, FEMA, Washington, DC <http://www.fema.gov/media-library/assets/documents/3522?id=1723> (October 6, 2014).

FEMA. (2008a). "Design and construction guidance for breakaway walls below elevated buildings in coastal high hazard areas." *Technical Bulletin 9*, FEMA, Washington, DC <http://www.fema.gov/media-library/assets/documents/3514?id=1722> (October 6, 2014).

FEMA. (2008b). "Flood damage-resistant materials requirements for buildings located in special flood hazard areas." *Technical Bulletin 2*, FEMA, Washington, DC <http://www.fema.gov/media-library/assets/documents/2655?id=1580> (October 6, 2014).

FEMA. (2008c). "Free-of-obstruction requirements for buildings located in coastal high hazard areas." *Technical Bulletin 5*, FEMA, Washington, DC <http://www.fema.gov/media-library/assets/documents/3490?id=1718> (October 6, 2014).

FEMA. (2008d). "Openings in foundation walls and walls of enclosures below buildings located in special flood hazard areas." *Technical Bulletin 1*, FEMA, Washington, DC <http://www.fema.gov/media-library/assets/documents/2644?id=1579> (October 6, 2014).

FEMA. (2008e). "Procedure memorandum 50: Policies and procedures for identifying and mapping areas subject to wave heights greater than 1.4 feet as an informational layer on flood insurance rate maps." FEMA, Washington, DC <http://www.fema.gov/media-library/assets/documents/34953> (October 6, 2014).

FEMA. (2009a). "Homeowner's guide to retrofitting: Six ways to protect your house from flooding." *FEMA P-312*, FEMA, Washington, DC <http://www.fema.gov/library/viewRecord.do?id=1420> (October 6, 2014).

FEMA. (2009b). "Protecting manufactured homes from floods and other hazards." *FEMA P-85*, FEMA, Washington, DC <http://www.fema.gov/media-library/assets/documents/2574> (October 6, 2014).

FEMA. (2010a). "Elevator installation for buildings located in special flood hazard areas." *Technical Bulletin 4*, FEMA, Washington, DC <http://www.fema.gov/media-library/assets/documents/3478> (October 6, 2014).

FEMA. (2010b). "Home builder's guide to coastal construction: Technical fact sheet series." *FEMA P-499*, FEMA, Washington, DC <http://www.fema.gov/library/viewRecord.do?id=2138> (October 6, 2013).

FEMA. (2010c). "Substantial improvement/substantial damage desk reference." *FEMA P-758*, FEMA, Washington, DC <http://www.fema.gov/media-library/assets/documents/18562?id=4160> (October 6, 2014).

FEMA. (2011). "Coastal construction manual: Principles and practices of planning, siting, designing, constructing, and maintaining residential buildings in coastal areas," 4th Ed. *FEMA P-55*, FEMA, Washington, DC <http://www.fema.gov/library/viewRecord.do?id=1671> (October 6, 2014).

FEMA. (2012a). "Engineering principles and practices for retrofitting flood prone residential buildings." *FEMA P-259*, FEMA, Washington, DC <http://www.fema.gov/library/viewRecord.do?id=1645> (October 6, 2014).

FEMA. (2012b). "Floodproofing certificate." *FEMA Form 086-0-34*, FEMA, Washington, DC <http://www.fema.gov/library/viewRecord.do?id=1600> (October 6, 2014).

FEMA. (2013a). "Floodproofing non-residential buildings." *FEMA P-936*, FEMA, Washington, DC <http://www.fema.gov/media-library/assets/documents/34270> (October 6, 2014).

FEMA. (2013b). "Operating guidance for improving the identification and mapping of the limit of moderate wave action (LiMWA) on regulatory and non-regulatory NFIP products." *Operating Guidance 13-13*, FEMA, Washington, DC <http://www.fema.gov/media-library/assets/documents/34953> (October 6, 2014).

Federal Highway Administration (FHA). (2012). "Hydraulic engineering circular 18: Evaluating scour at bridges." *FHWA-HIF 12-003*, FHA, Washington, DC.

Forest Products Laboratory. (2010). "Wood handbook—Wood as an engineering material." *General Technical Rep. FPL-GTR-190*. U.S. Department of Agriculture, Forest Service, Forest Products Laboratory, Madison, WI.

Fowler, J. E. (1993). "Coastal scour problems and prediction of maximum scour." *Technical Report CERC-93-8*, U.S. Army Corps of Engineers, Washington, DC <http://chl.erdc.usace.army.mil/chl.aspx?p=s&a=Publications;399> (October 6, 2014).

Jones, C. P., Coulbourne, W. L., and Tertell, P. (2001). "Consideration of a new flood hazard zone: The coastal A zone." *Proc., Association of State Floodplain Managers Annual Conference*, Charlotte, NC.

National Academy of Sciences (NAS). (1977). *Methodology for calculating wave action effects associated with storm surges*. NAS, Washington, DC.

National Weather Service, Hydrologic Services Program (NWS/HSP). (2012). *Flood warning systems manual*. NWSPD, Silver Spring, MD <http://www.nws.noaa.gov/om/water/ahps/resources/Flood_Warning-Systems_Manual.pdf> (October 6, 2014).

U.S. Army Corps of Engineers (USACE). (1975). *Guidelines for identifying coastal high hazard zones*. Galveston District Corps of Engineers, Galveston, TX.

USACE. (1995). "Flood proofing regulations." *EP 1165-2-314*. USACE, Washington, DC <http://www.usace.army.mil/Portals/2/docs/civilworks/Project%20Planning/nfpc/Flood%20Proofing%20Regulations%201995.pdf> (October 6, 2014).

USACE. (2000). "Engineering and design: Design and construction of levees." *Engineer Manual No. 1110-2-1913*. USACE, Washington, DC <http://www.publications.usace.army.mil/Portals/76/Publications/EngineerManuals/EM_1110-2-1913.pdf> (October 6, 2014).

USACE. (2002). "Ice engineering." *Engineer Manual No. 1110-2-1612*. USACE, Washington, DC <http://www.publications.usace.army.mil/Portals/76/Publications/EngineerManuals/EM_1110-2-1612.pdf> (October 6, 2014).

USACE. (2004). "General design and construction consideration for earth and rock-fill dams." *Engineer Manual No. 1110-2-2300*. USACE, Washington, DC <http://www.publications.usace.army.mil/Portals/76/Publications/EngineerManuals/EM_1110-2-2300.pdf> (October 6, 2014).

USACE. (2006). *Coastal engineering manual*. USACE, Washington, DC <http://chl.erdc.usace.army.mil/cem> (October 6, 2014).

INDEX

Page numbers followed by *e*, *f*, and *t* indicate equations, figures, and tables, respectively.